Combustion

Combustion

IRVIN GLASSMAN

Department of Aerospace and Mechanical Sciences
and Center for Environmental Studies
Princeton University
Princeton, New Jersey

ACADEMIC PRESS New York San Francisco London 1977

A Subsidiary of Harcourt Brace Jovanovich, Publishers

ACADEMIC PRESS, INC.
111 Fifth Avenue, New York, New York 10003

United Kingdom Edition published by
ACADEMIC PRESS, INC. (LONDON) LTD.
24/28 Oval Road, London NW1

Library of Congress Cataloging in Publication Data

Glassman, Irvin.
Combustion.

Includes bibliographies.
1. Combustion. I. Title.
QD516.G55 541'.361 76-27442
ISBN: 0-12-285850-6

PRINTED IN THE UNITED STATES OF AMERICA

To

BEV

who was very much a part of all of this,

on

Our 25th Wedding Anniversary

"No man can reveal to you aught but that which already lies half asleep in the dawning of your knowledge.

If he (the teacher) is wise he does not bid you to enter the house of his wisdom, but leads you to the threshold of your own mind.

The astronomer may speak to you of his understanding of space, but he cannot give you his understanding.

And he who is versed in the science of numbers can tell of the regions of weight and measures, but he cannot conduct you thither.

For the vision of one man lends not its wings to another man."

Gibran, THE PROPHET

Contents

Chapter Four **Flame Phenomena in Premixed Combustible Gases**

Chapter Five **Detonation**

Preface

During my twenty years of teaching combustion at Princeton, I had accumulated extensive lecture notes and developed my own approach to the subject matter. For years former students and associates have encouraged me to publish these notes. This book is the result.

My whole concept of teaching has been to stimulate the student to think, to learn the material on his own, and to understand how to use it in his own research and development endeavors. It is difficult to assess whether this concept will prevail in this book. Combustion is a most complex subject that involves primarily the disciplines of chemistry, physics, and fluid mechanics. However, it is important to understand that approaches to a complex subject can be made in a fundamental manner. One must gain the physical insight into the underlying principles. Although many subjects are presented, I have tried to strip away the complexities and elaborate upon the physical insights essential to understanding. Chapter IX on coal combustion epitomizes this approach. When I thought it necessary to cover this topic in class, I was surprised that there was not readily available in the literature some of the simple results developed in this chapter.

The subject matter which comprises the field of combustion is diverse. No attempt has been made to develop a unified approach to all the material. Indeed, in my opinion, in order to gain the best insight the approach should vary with the subject matter.

Acknowledgments

My understanding of combustion came about from many associations. The one that I have cherished the most has been with my own graduate students. Their contributions to this book are many. In particular, I must recognize and thank Dr. F. L. Dryer who chose to remain at Princeton and assume numerous responsibilities in our laboratory while I undertook other endeavors—such as writing this book.

The foundation for much of what I have written was developed during 25 years of research in the field. I had no previous experience or training in this area prior to coming to Princeton. Practically all my Princeton research was sponsored by the U.S. government. Thus I would also like to recognize the confidence expressed in me by the technical monitors of my research contracts and grants. They deserve the thanks of many of us. In particular, I owe much to Dr. J. F. Masi of the Air Force Office of Scientific Research for his particular interest in the contributions he thought I could make by my approach to combustion. I hope this book is another such contribution.

Special thanks are due to my wife and children who gave me the love and happiness necessary to pursue this arduous, but enjoyable, career.

Chapter One

Chemical Thermodynamics

The most essential parameters necessary for the evaluation of combustion systems are the equilibrium product temperature and composition. If all the heat evolved in the reaction is employed solely to raise the product temperature, then this temperature is called the adiabatic flame temperature. Because of the importance of the temperature and gas composition in combustion considerations, it is useful to review those aspects of the field of chemical thermodynamics which deal with these subjects.

A. HEATS OF REACTION AND FORMATION

All chemical reactions are accompanied either by an absorption or an evolution of energy, which usually manifests itself as heat. It is possible to determine this amount of heat and thus the temperature and product composition from very basic principles. Statistical calculations permit one to determine the internal energy of a substance. The internal energy of a given substance is found to be dependent upon its temperature, pressure, and state and is independent of the means by which the state was attained. Likewise the change in internal energy ΔE of a system which results from any physical change or chemical reaction depends only on the initial and final state of the system. The total change in internal energy will be the same, whether or not the energy is evolved in any form of heat, energy, or work.

For a flow reaction proceeding with negligible changes in kinetic energy, potential energy, and with no form of work beyond that required for flow, the heat added is equal to increase of enthalpy of the system

$$Q = \Delta H$$

For a nonflow reaction proceeding at constant pressure, the heat added is also equal to the gain in enthalpy

$$Q_P = \Delta H$$

and if heat is evolved

$$Q_P = -\Delta H$$

Most thermochemical calculations are made for closed thermodynamic systems; the stoichiometry is most conveniently represented in terms of the molar quantities as determined from statistical calculations. In compressible flow problems in which open thermodynamic systems are essential, it is best to deal with mass quantities. Upper case symbols will be used for molar quantities, and lower case systems will be used for mass quantities.

One of the most important thermodynamic facts to be known about a given chemical reaction is the change in energy or heat content associated with the reaction at some specified temperature with each of the reactants and products in an appropriate standard rate. This change is known either as the energy or heat of reaction at the specified temperature.

The standard state means that for each state a reference state of the aggregate exists. For gases, the thermodynamic standard reference state is taken to be equal to that of the ideal gaseous state at atmospheric pressure at each temperature. The ideal gaseous state is the case of isolated molecules which give no interactions and which obey the equation of state of a perfect gas. The standard reference state for pure liquids and solids at a given temperature is the real state of the substance at a pressure of one atmosphere.

The thermodynamic symbol which represents the property of the substance in the standard state at a given temperature is written, for example, as $H_T{}^\circ$, $E_T{}^\circ$, etc., where the superscript $^\circ$ specifies the standard state and the subscript T the specific temperature. Statistical calculations actually permit the determination of $E_T - E_0$, which is the energy content at a given temperature referred to the energy content at 0°K. For one mole in the ideal gaseous state,

$$PV = RT \tag{1}$$

$$H^\circ = E^\circ + (PV)^\circ = E^\circ + RT \tag{2}$$

which at 0°K reduces to

$$H_0^\circ = E_0^\circ \tag{3}$$

Thus the heat content at any temperature referred to the heat or energy content at 0°K is known and is

$$(H^\circ - H_0^\circ) = (E^\circ - E_0^\circ) + RT \tag{4}$$

From the definition of the heat of reaction, Q_P will depend on the temperature T at which the reaction and product enthalpies are evaluated. The heat of reaction at one temperature T_0 can be related to that at another temperature T_1. Consider the reaction configuration shown in Fig. 1.

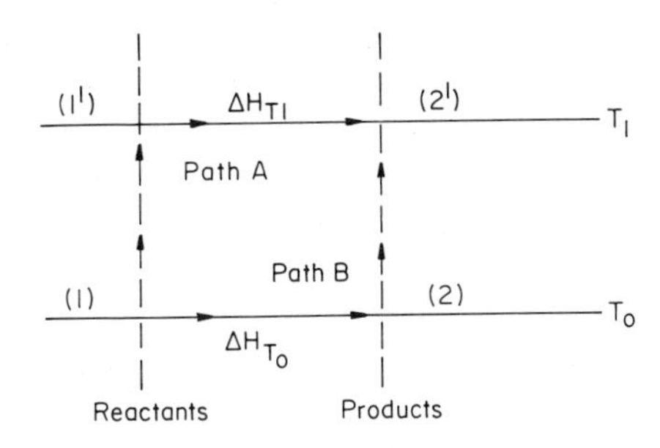

Fig. 1. Temperature-reaction paths.

According to the First Law, the heat changes which proceed from reactants at temperature T_0 to products at temperature T_1 by either Path A or Path B shown must be the same. Path A raises the reactants from temperature T_0 to T_1 and reacts at T_1. Path B reacts at T_0 and raises the products from T_0 to T_1. The energy equality which relates the heats of reaction at the two different temperatures is written as

$$\sum_{j\,\text{react}} n_j[(H_{T_1}^\circ - H_0^\circ) - (H_{T_0}^\circ - H_0^\circ)]_j + \Delta H_{T_1} = \Delta H_{T_0} + \sum_{i\,\text{prod}} n_i[(H_{T_1}^\circ - H_0^\circ) - (H_{T_0}^\circ - H_0^\circ)]_i \tag{5}$$

Any changes of phase can be included in the heat content terms. Thus by knowing the difference in energy content at the different temperatures for the products and reactants, it is possible to determine the heat of reaction at one temperature from the heat of reaction at another.

If the heats of reaction at a given temperature are known for two separate reactions, the heat of reaction of a third reaction at the same temperature may be determined by simple algebraic addition. This statement is the Law of Heat Summation. For example, two reactions which can be carried out conveniently in a calorimeter at constant pressure are

$$C_{graphite} + O_2(g) \xrightarrow[298]{} CO_2(g), \quad Q_P = +94.1 \quad kcal \tag{6}$$

$$CO(g) + \tfrac{1}{2}O_2(g) \xrightarrow[298]{} CO_2(g), \quad Q_P = +67.7 \quad kcal \tag{7}$$

Subtracting these two reactions, one obtains

$$C_{graphite} + \tfrac{1}{2}O_2(g) \xrightarrow[298]{} CO(g), \quad Q_P = +26.4 \quad kcal \tag{8}$$

Since some of the carbon would burn to CO_2 and not just CO, it is difficult to determine calorimetrically the heat released by reaction (8).

It is, of course, not necessary to have an extensive list of heats of reaction to determine the heat absorbed or evolved in every possible chemical reaction. A more convenient and logical procedure is to list what are known as the standard heats of formation of chemical substances. The standard heat of formation is the enthalpy of a substance in its standard state referred to its elements in their standard states at the same temperature. From this definition it is obvious that the heats of formation of the elements in their standard states are zero.

The value of the heat of formation of a given substance from its elements may be the result of the determination of the heat of one reaction. Thus from the calorimeteric reaction for burning carbon to CO_2 (Eq. (6)), it is possible to write the heat of formation of carbon dioxide at 298°K as

$$(\Delta H_f^\circ)_{298, CO_2} = -94.1 \quad kcal/mole$$

The superscript to the heat of formation symbol ΔH_f represents the standard state and the subscript number the base or reference temperature. From the example for the Law of Heat Summation, it is apparent that the heat of formation of carbon monoxide from Eq. (8) is

$$(\Delta H_f^\circ)_{298, CO} = -26.4 \quad kcal/mole$$

It is evident that, by judicious choice, the number of reactions which must be measured calorimetrically will be about the same as the number of substances whose heats of formation are to be determined.

The logical consequence of the above is that given the heats of formation of the substances which make up any particular reaction, one can determine directly the heat of reaction or heat evolved at the reference temperature as follows

$$\Delta H_T = \sum_{i\ prod} n_i(\Delta H_f^\circ)_{T,i} - \sum_{j\ react} n_j(\Delta H_f^\circ)_{T,j} = -Q_p \tag{9}$$

There exist extensive tables of standard heats of formation; however, all are not at the same reference temperature. The most convenient are the

TABLE 1

Heats of formation at 298.1°K

Chemical symbol	Name	State	ΔH_f° (kcal/mole)
C	Carbon	Vapor	126.36
N	Nitrogen atom	Gas	112.75
O	Oxygen atom	Gas	59.16
C_2H_2	Acetylene	Gas	54.19
H	Hydrogen atom	Gas	52.09
O_3	Ozone	Gas	34.00
NO	Nitric oxide	Gas	21.60
C_6H_6	Benzene	Gas	19.80
C_6H_6	Benzene	Liquid	11.71
C_2H_4	Ethane	Gas	12.50
N_2H_4	Hydrazine	Liquid	12.05
OH	Hydroxyl radical	Gas	10.06
O_2	Oxygen	Gas	0
N_2	Nitrogen	Gas	0
H_2	Hydrogen	Gas	0
C	Carbon	Solid	0
NH_3	Ammonia	Gas	−11.04
C_2H_4O	Ethylene oxide	Gas	−12.19
CH_4	Methane	Gas	−17.89
C_2H_4	Ethane	Gas	−20.24
CO	Carbon monoxide	Gas	−26.42
C_4H_{10}	Butane	Gas	−29.81
CH_3OH	Methanol	Gas	−48.10
CH_3OH	Methanol	Liquid	−57.04
H_2O	Water	Gas	−57.80
C_8H_{18}	Octane	Liquid	−59.74
H_2O	Water	Liquid	−68.32
SO_2	Sulfur dioxide	Gas	−71.00
$C_{12}H_{16}$	Dodecane	Liquid	−83.00
CO_2	Carbon dioxide	Gas	−94.05
SO_3	Sulfur trioxide	Gas	−94.45

compilations known as the JANNAF and NBS Tables, both of which use 298°K as the reference temperature. Table 1 lists some values of the heat of formation taken from Penner (1957), NBS Circular C461 (1947), and other sources.

B. FREE ENERGY, THE EQUILIBRIUM CONSTANT, AND FLAME TEMPERATURE CALCULATIONS

For those cases in which the products are measured at a different temperature T_2 than the reference temperature T_0, and the reactants enter the system at a different temperature T_1 than the reference temperature, the heat of reaction becomes

$$\begin{aligned}\Delta H_R = &\sum_{i\,\text{prod}} n_i[\{(H^\circ_{T_2} - H_0^\circ) - (H^\circ_{T_0} - H_0^\circ)\} + (\Delta H_f^\circ)_{T_0}]_i \\ &- \sum_{j\,\text{react}} n_j[\{(H^\circ_{T_1} - H_0^\circ) - (H^\circ_{T_0} - H_0^\circ)\} + (\Delta H_f^\circ)_{T_0}]_j \\ &= -Q_P \quad \text{(evolved)}\end{aligned} \tag{10}$$

Most systems are considered to have the reactants enter at the standard reference temperature 298°K. Consequently the enthalpy terms in the braces for the reactants disappear. The JANNAF Tables tabulate, as a supposed convenience, $(H_T^\circ - H^\circ_{298})$ instead of $(H_T^\circ - H_0^\circ)$. This type of tabulation is unfortunate since for systems using cryogenic fuels and oxidizers, such as rockets, the reactants can enter the system at temperatures lower than the reference temperature. Indeed the fuel and oxidizer could enter at different temperatures, and the summation in Eq. (10) can be handled conveniently by first realizing that T_1 may vary with the substance j.

When all the heat evolved is used to heat the product gases, the product temperature T_2 is called the adiabatic flame temperature. In this case $\Delta H_R = 0$ and Eq. (10) becomes

$$\begin{aligned}&\sum_{i\,\text{prod}} n_i[\{(H^\circ_{T_2} - H_0^\circ) - (H^\circ_{T_0} - H_0^\circ)\} + (\Delta H_f^\circ)_{T_0}]_i \\ &\quad = \sum_{j\,\text{react}} n_j[\{(H^\circ_{T_1} - H_0^\circ) - (H^\circ_{T_0} - H_0^\circ)\} + (\Delta H_f^\circ)_{T_0}]_j\end{aligned} \tag{11}$$

If the products n_i of this reaction are known, then Eq. (11) can be solved for the flame temperature. For a reacting system whose product temperature is less than 1250°K, the products are the normal stable species CO_2, H_2O, N_2, and O_2, whose molar quantities can be determined from simple mass balances. However, most combustion systems reach temperatures appreciably greater than 1250°K and dissociation of the stable species occurs. Since the dissociation reactions are quite endothermic, a few percent dissociation can lower the flame temperature substantially. The stable products from a C–H–O reaction system can dissociate by any of the following reactions:

$$CO_2 \rightleftharpoons CO + \tfrac{1}{2}O_2$$

$$CO_2 + H_2 \rightleftharpoons CO + H_2O$$

$$H_2O \rightleftharpoons H_2 + \tfrac{1}{2}O_2$$

$$H_2O \rightleftharpoons H + OH$$

$$H_2O \rightleftharpoons \tfrac{1}{2}H_2 + OH$$

$$H_2 \rightleftharpoons 2H$$

$$O_2 \rightleftharpoons 2O, \quad \text{etc.}$$

Each of these reactions helps specify a definite equilibrium concentration of each product at a given temperature. Whereas in heat of reaction experiments or low temperature combustion experiments, the products could be specified from the chemical stoichiometry, ones sees now that with dissociation the specification of the product concentrations becomes much more complex and the n_i's in the flame temperature equation [Eq. (11)] are as unknown as the flame temperature itself. In order to solve the equation for the n_i's and T_2, it is apparent that more than mass balance equations are needed. The necessary equations are found in the equilibrium relationships which exist among the product composition in the equilibrium system.

Recall from elementary thermodynamics that the condition for equilibrium of a chemical system comes from the following form of the Second Law

$$T\,dS = dE + P\,dV \tag{12}$$

At constant temperature and pressure, Eq. (12) takes the form

$$d(E + PV - TS)_{T,P} = 0 \tag{13}$$

or

$$d(H - TS)_{T,P} = 0 \tag{14}$$

By definition, the free energy F is written as

$$F \equiv E + PV - TS = H - TS \tag{15}$$

Thus the condition for equilibrium at constant temperature and pressure is that the change in free energy be zero, i.e.,

$$(dF)_{T,P} = 0 \tag{16}$$

If one considers an arbitrary equilibrium reaction

$$a\mathrm{A} + b\mathrm{B} \rightleftharpoons r\mathrm{R} + s\mathrm{S} \tag{17}$$

it can be shown that for ideal gases,

$$\Delta F^\circ = -RT\ \ln(p_\mathrm{R}{}^r p_\mathrm{S}{}^s / p_\mathrm{A}{}^a p_\mathrm{B}{}^b) \tag{18}$$

where the pressures are measured in atmospheres. The logarithm of the pressures arises from the relationship of the entropy to the pressure. One can define

$$K_P \equiv p_\mathrm{R}{}^r p_\mathrm{S}{}^s / p_\mathrm{A}{}^a p_\mathrm{B}{}^b \tag{19}$$

as the proper quotient of partial pressure at equilibrium. K_P is called the equilibrium constant at constant pressure and K_P is not a function of total pressure but of temperature alone. This statement is clear since ΔF° is a function of temperature alone. It is a little surprising that the free energy change at the standard state pressure (1 atm) determines the equilibrium condition at all other pressures. Equations (11) and (19) can be modified to account for nonideality in the product state; however, because of the high temperatures reached in combustion systems, ideality can be assumed even under rocket chamber pressures.

The flame temperature equation is more conveniently written in terms of moles, and thus it is best to write the K_P in terms of moles. This conversion is accomplished through the relationship between the partial pressure p and the total pressure P.

$$p_i = x_i P \tag{20}$$

where x_i is the mole fraction and is defined as

$$x_i = n_i / \sum n_i \tag{21}$$

and Eq. (20) becomes

$$p_i = (n_i / \sum n_i) P \tag{22}$$

Substituting this expression for p_i in the definition of the equilibrium constant (Eq. (19)), one obtains

$$K_P = (n_R{}^r n_S{}^s / n_A{}^a n_B{}^b)(P / \sum n_i)^{r+s-a-b} \tag{23}$$

which is sometimes written

$$K_P = K_N (P / \sum n_i)^{r+s-a-b} \tag{24}$$

where

$$K_N \equiv n_R{}^r n_S{}^s / n_A{}^a n_B{}^b \tag{25}$$

When

$$r + s - a - b = 0 \tag{26}$$

the equilibrium reaction is said to be pressure insensitive. However, it is worth repeating that K_P is not a function of pressure, but Eq. (24) shows then that K_N can be a function of pressure.

The equilibrium constant based on concentration (moles/cm^3) is sometimes used, particularly in chemical kinetic analyses to be discussed in the next chapter. This constant is found by recalling that the perfect gas law states

$$PV = \sum n_i RT \tag{27}$$

or

$$(P / \sum n_i) = (RT / V) \tag{28}$$

where V is the volume. Substituting for $(P/\sum n_i)$ in Eq. (24) gives

$$K_P = (n_R{}^r n_S{}^s / n_A{}^a n_B{}^b)(RT/V)^{r+s-a-b} \tag{29}$$

or

$$K_P = \frac{(n_R/V)^r (n_S/V)^s}{(n_A/V)^a (n_B/V)^b} (RT)^{r+s-a-b} \tag{30}$$

Equation (30) can be written as

$$K_P = (C_R{}^r C_S{}^s / C_A{}^a C_B{}^b)(RT)^{r+s-a-b} \qquad \text{where} \quad C = n/V \tag{31}$$

From Eq. (31) it is seen that the definition of the equilibrium constant for concentrations is

$$K_C \equiv C_R{}^r C_S{}^s / C_A{}^a C_B{}^b \tag{32}$$

Given a temperature and pressure all the equilibrium constants (K_P, K_N, and K_C) can be determined thermodynamically from ΔF° for the equilibrium reaction chosen. The mean molecular weight is also necessary to determine K_N.

In the same context as the heats of formation are listed, the JANNAF Tables have tabulated the equilibrium constants of formation for practically every substance of concern in combustion systems. The equilibrium constant of formation is based on the equilibrium equation of formation of a species from its elements in their normal states. Thus by algebraic manipulation it is possible to determine the equilibrium constant of any reaction. In flame temperature calculations, by dealing only with the equilibrium systems of formation reactions, there is no chance of choosing a redundant set of equilibrium reactions. Again, the equilibrium constant of formation for the elements in their normal states is unity.

Consider the following three equilibrium reactions of formation:

$$H_2 + \tfrac{1}{2}O_2 \rightleftharpoons H_2O, \qquad K_{P,\,f(H_2O)} = p_{H_2O}/p_{H_2}\,p_{O_2}^{1/2}$$

$$\tfrac{1}{2}H_2 \rightleftharpoons H, \qquad K_{P,\,f(H)} = p_H/p_{H_2}^{1/2}$$

$$\tfrac{1}{2}O_2 + \tfrac{1}{2}H_2 \rightleftharpoons OH, \qquad K_{P,\,f(OH)} = p_{OH}/p_{O_2}^{1/2}\,p_{H_2}^{1/2}$$

The equilibrium reaction is always written for the formation of one mole of the substances other than the elements. Now, if one desires to calculate the equilibrium reaction for a reaction such as

$$H_2O \rightleftharpoons H + OH$$

the K_P is then found to be

$$K_P = p_H\, p_{OH}/p_{H_2O} = K_{Pf(H)}\, K_{Pf(OH)}/K_{Pf(H_2O)}$$

Because of this type of result and the thermodynamic result

$$\Delta F^\circ = -RT \ln K_P$$

the JANNAF Tables list "ln $K_{P,\mathrm{f}}$."

For those compounds that contain carbon and for combustion systems in which solid carbon is formed, the thermodynamic handling of the K_P is somewhat more difficult. The equilibrium reaction for CO_2 would be

$$C_{\mathrm{graphite}} + O_2 \rightleftharpoons CO_2, \qquad K_P = p_{CO_2}/p_{O_2} p_C$$

However, since the carbon is in the condensed state, the only partial pressure it exerts is its vapor pressure, a known thermodynamic property, which is also a function of temperature. Thus the above formation expression is written as

$$K_P(T) P_{vp,\mathrm{C}}(T) = p_{CO_2}/p_{O_2} = K_P'$$

The $K_{P,\mathrm{f}}$'s for substances containing carbon tabulated by JANNAF are in reality K_P', and the condensed phase is simply ignored in the equilibrium expression. Carbon or any other condensed phase is not included in the $\sum n_i$ since this summation is for the gas phase components contributing to the total pressure.

If one examines the equation for the flame temperature (Eq. (11)), an interesting deduction can be made. The values in Table 1 and the realization that many moles of product form for each mole of the reactant fuel will show that the sum of the molar heats of the products will be substantially greater than the sum of the molar heats of the reactants; i.e., $\sum_{i\,\mathrm{prod}} n_i(\Delta H_{\mathrm{f}}{}^\circ)_i \gg \sum_{j\,\mathrm{react}} n_i(\Delta H_{\mathrm{f}}{}^\circ)_j$. Consequently one can conclude that the flame temperature is not determined by the specific reactants, but only by the atomic ratios and the specific atoms that are introduced. It is the atoms which determine which products will form. Only ozone and acetylene have high enough positive heats of formation to cause a noticeable variation (rise) in flame temperature. Ammonia has a low enough negative heat of formation to lower the final flame temperature. Table 1, which lists the heats of formation of various compounds, is very much like a potential energy diagram in which movement from the top of the table to substances below indicates energy release.

One can draw the further conclusion that the product concentrations also are only functions of temperature and pressure and not the original source of atoms. Thus for any C—H—O system, the products will be the same, i.e., they will be CO_2, H_2O, and their dissociated products. The dissociation reactions listed earlier give some of the possible "new" products. A more complete list would be

$$CO_2, \quad H_2O, \quad CO, \quad H_2, \quad O_2, \quad OH, \quad H, \quad O, \quad O_3, \quad C, \quad CH_4$$

For a C, H, O, N system, the following should be added:

$$N_2, \quad N, \quad NO, \quad NH_3, \quad NO^+, \quad e$$

NO has a very low ionization potential and will ionize at flame temperatures. For a normal composite solid propellant which contains C, H, O, N, Cl, and Al, many more products would have to be considered. In fact if all the possible number of products for this system were listed the solution to the problem would tax some of the more advanced computers. However, knowledge of thermodynamic equilibrium constants and kinetics allows one to eliminate many possible product species.

Consider a C, H, O, N system. For an overoxidized case, there is an excess of oxygen to convert all the carbon and hydrogen present to CO_2 and H_2O. The term stoichiometric mixture ratio specifies exactly the ratio of the oxygen to fuel to just burn all the C and H to CO_2 and H_2O. The principal products for an overoxidized case are CO_2, H_2O, O_2, and N_2. As the temperature of the flame increases, dissociation begins; and if $T_2 > 2200°K$ at $P = 1$ atm or $T_2 > 2500°K$ at 20 atm, the dissociation of CO_2 and H_2O must be accounted for by the reactions

$$CO_2 \rightleftharpoons CO + \tfrac{1}{2}O_2, \qquad Q_P = -27.8 \text{ kcal}$$

$$H_2O \rightleftharpoons H_2 + \tfrac{1}{2}O_2, \qquad Q_P = -57.8 \text{ kcal}$$

$$H_2O \rightleftharpoons \tfrac{1}{2}H_2 + OH, \qquad Q_P = -67.1 \text{ kcal}$$

This heuristic rule is based upon the fact that under these temperatures and pressures at least 1% dissociation takes place. The pressure effect enters through Le Chatelier's principle that the equilibrium concentrations will shift with the pressure. The equilibrium constant, although independent of pressure, was expressed in a form which contained the pressure. A variation in pressure shows that the molar quantities change. Since the reactions noted are quite endothermic, even small concentrations must be considered. If one initially assumes that certain products of dissociation were absent and calculates a temperature which indicates about 1% of such products, then the flame temperature must be reevaluated by including in the product mixture these products of dissociation, i.e., the presence of CO, H_2, OH as products is now indicated.

Since problems about emissions from power plant sources have arisen, the concentration of certain products much less than 1% must be considered, even though such concentrations do not affect the temperature even in a minute way. The major pollutant in this regard is NO. To make an estimate of the amount of NO found in a system at equilibrium, the equilibrium reaction of formation of NO is used

$$\tfrac{1}{2}N_2 + \tfrac{1}{2}O_2 \rightleftharpoons NO$$

A rule of thumb is that any temperature above 1800°K gives sufficient NO to be of concern. The NO formation reaction is pressure insensitive, so there is no need to specify the pressure.

If in the overoxidized case $T_2 > 2400°K$ at $P = 1$ atm and $T_2 > 2800°K$ at $P = 20$ atm, then the dissociation of O_2 and H_2 becomes important, viz.,

$$H_2 \rightleftharpoons 2H, \qquad Q_P = -103.8 \quad \text{kcal}$$

$$O_2 \rightleftharpoons 2O, \qquad Q_P = -117.2 \quad \text{kcal}$$

Although these dissociation reactions are written to show 1 mole of the molecule dissociating, recall that the $K_{P,f}$'s are written for 1 mole of the radical forming. These dissociation reactions are highly endothermic, and even very small percentages can affect the final temperature. The new products are H and O atoms. Actually the presence of O atoms could come about from the dissociation of water at this higher temperature according to the equilibrium step

$$H_2O \rightleftharpoons H_2 + O, \qquad Q_P = -116.9 \quad \text{kcal}$$

From Le Chatelier's principle there is basically no preference in the reactions leading to O since the heat absorption is about the same in each case. Thus in an overoxidized flame, water dissociation introduces the species H_2, O_2, OH, H, and O.

At even higher temperatures, the nitrogen begins to take part in the reactions and to affect the system thermodynamically. At $T > 3000°K$, NO forms mostly from the reaction given before:

$$\tfrac{1}{2}N_2 + \tfrac{1}{2}O_2 \rightleftharpoons NO, \qquad Q_P = -21.5 \quad \text{kcal}$$

rather than from

$$\tfrac{1}{2}N_2 + H_2O \rightleftharpoons NO + H_2, \qquad Q_P = -79.3 \quad \text{kcal}$$

If $T_2 > 3500°K$ at $P = 1$ atm or $T > 3600°K$ at 20 atm, N_2 starts to dissociate

$$N_2 \rightleftharpoons 2N, \qquad Q_P = -225.1 \quad \text{kcal}$$

another highly endothermic reaction.

Thus the complexity in solving for the flame temperature depends on the number of product species chosen. If the approximate temperature range of a system is known, then by the approach indicated above, the complexity of the system can be reduced. Unfortunately, computer programs and machines are now available that can handle the most complex systems and little thought is given to reducing the complexity and thus the machine time.

Equation (11) is now examined more closely. If the n_i's (products) total a number μ, one needs $(\mu + 1)$ equations to solve for the μn_i's and

T_2. The energy equation is available as one equation. Further, there is a mass balance equation for each atom in the system. If there are α atoms, then $(\mu - \alpha)$ additional equations are required to solve the problem. The $(\mu - \alpha)$ equations formed from the equilibrium equations are basically nonlinear. For the CHON system it is necessary to solve five linear equations and $(\mu - 4)$ nonlinear equations simultaneously, in which one of the unknowns T_2 is present in terms of the enthalpies of the products only. This set of equations is difficult to solve and can be done only with modern computational machinery.

Consider the reaction between octane and nitric acid taking place at a pressure P as an example. The stoichiometric equation is written as

$$n_{C_8H_{18}}C_8H_{18} + n_{HNO_3}HNO_3 \rightleftharpoons n_{CO_2}CO_2 + n_{H_2O}H_2O + n_{H_2}H_2 + n_{CO}CO + n_{O_2}O_2 + n_{N_2}N_2 + n_{OH}OH + n_{NO}NO + n_OO + n_CC_{solid} + n_HH$$

Since the mixture ratio is not specified explicitly for this general expression, no effort is made to eliminate products $\mu = 11$. The new mass balance equations then are $(\alpha = 4)$

$$N_H = 2n_{H_2} + 2n_{H_2O} + n_{OH} + n_H$$

$$N_O = 2n_{O_2} + n_{H_2O} + 2n_{CO_2} + n_{CO} + n_{OH} + n_O + n_{NO}$$

$$N_N = 2n_{N_2} + n_{NO}, \qquad N_C = n_{CO_2} + n_{CO} + n_C$$

where

$$N_H = 18n_{C_8H_{18}} + n_{HNO_3}, \qquad N_O = 3n_{HNO_3}, \qquad N_C = 8n_{C_8H_{18}}, \qquad N_N = n_{HNO_3}$$

The seven $(\mu - \alpha = 11 - 4 = 7)$ equilibrium equations needed would be

$$C + O_2 \rightleftharpoons CO_2, \qquad K_{P,f} = n_{CO_2}/n_{O_2} \tag{i}$$

$$H_2 + \tfrac{1}{2}O_2 \rightleftharpoons H_2O, \qquad K_{P,f} = (n_{H_2O}/n_{H_2}n_{O_2}^{1/2})(P/\textstyle\sum n_i)^{-1/2} \tag{ii}$$

$$C + \tfrac{1}{2}O_2 \rightleftharpoons CO, \qquad K_{P,f} = (n_{CO}/n_{O_2}^{1/2})(P/\textstyle\sum n_i)^{1/2} \tag{iii}$$

$$\tfrac{1}{2}H_2 + \tfrac{1}{2}O_2 \rightleftharpoons OH, \qquad K_{P,f} = n_{OH}/n_{H_2}^{1/2}n_{O_2}^{1/2} \tag{iv}$$

$$\tfrac{1}{2}O_2 + \tfrac{1}{2}N_2 \rightleftharpoons NO, \qquad K_{P,f} = n_{NO}/n_{O_2}^{1/2}n_{N_2}^{1/2} \tag{v}$$

$$\tfrac{1}{2}O_2 \rightleftharpoons O, \qquad K_{P,f} = (n_O/n_{O_2}^{1/2})(P/\textstyle\sum n_i)^{1/2} \tag{vi}$$

$$\tfrac{1}{2}H_2 \rightleftharpoons H, \qquad K_{P,f} = (n_H/n_{H_2}^{1/2})(P/\textstyle\sum n_i)^{1/2} \tag{vii}$$

In these equations $\sum n_i$ includes only the gaseous products; i.e., it does not include n_C, which is determined from the equation for N_C.

The reaction between the reactants and products is considered nonreversible, so that only the products exist in the system being analyzed. Thus if the reactants were H_2 and O_2, both would appear on the product side as well. In dealing with the equilibrium reactions, the molar quantities of the reactants H_2 and O_2 are ignored. They are given or known quantities. The amounts of H_2 and O_2 in the product mixture would be unknowns. This point should be considered carefully, even though obvious. It is one of the major sources of error in first attempts to solve flame temperature problems.

There are various mathematical approaches for solving these equations by numerical methods. These are not discussed here. Glassman and Sawyer (1970) give some details in their book. Huff and Morell (1950) and Gordon and McBride (1971) discuss the methods of solution extensively.

In combustion calculations, it is the variation of the temperature with the oxidizer to fuel ratio that is desired. Therefore in solving flame temperature problems, it is normal to take the number of moles of fuel as 1 and the number of moles of oxidizer as that given by the oxidizer to fuel ratio. In this manner the coefficients are 1 and a number normally larger than 1. Plots of flame temperature versus oxidizer to fuel ratio peak at the stoichiometric mixture rate. If the system is overoxidized, there is excess oxygen which must be heated to the product temperature, and thus the product temperature drops from the stoichiometric value. If too little oxidizer is present, i.e., the system is underoxidized, then there is not sufficient oxygen to burn all the carbon and hydrogen to their most oxidized state, the energy release is less and the temperature drops as well. More generally the flame temperature is plotted as a function of equivalence ratio (Fig. 2), where the equivalence ratio is defined as the fuel/oxidizer

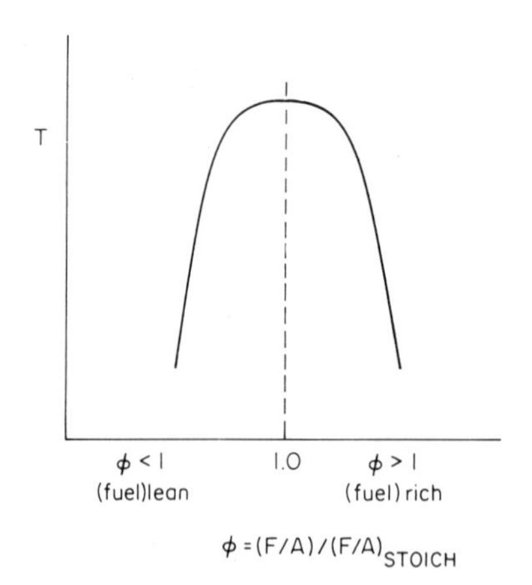

Fig. 2. Variation of temperature with equivalence ratio.

ratio divided by the stoichiometric fuel/oxidizer ratio. The equivalence ratio is given the symbol ϕ. For fuel-rich systems, there is more than the stoichiometric amount of fuel and $\phi > 1$. For overoxidized or fuel-lean systems, $\phi < 1$. Obviously at stoichiometric, $\phi = 1$.

In actuality some systems have their temperature peak slightly on the rich side of stoichiometric. This result occurs because, if the system is slightly underoxidized, the specific heat of the products is reduced and thus the flame temperature increased. Some maximum flame temperatures are given in Table 2.

TABLE 2

Approximate flame temperatures of various stoichiometric mixtures. Critical temperature 298°K

Fuel	Oxidizer	Pressure (atm)	T (°K)
Acetylene	Air	1	2600[a]
Acetylene	Oxygen	1	3410[b]
Carbon monoxide	Air	1	2400
Carbon monoxide	Oxygen	1	3220
Heptane	Air	1	2290
Heptane	Oxygen	1	3100
Hydrogen	Air	1	2400
Hydrogen	Oxygen	1	3080
Methane	Air	1	2210
Methane	Air	20	2270
Methane	Oxygen	1	3030
Methane	Oxygen	20	3460

[a] This maximum exists at $\phi = 1.3$.
[b] This maximum exists at $\phi = 1.7$.

REFERENCES

Glassman, I., and Sawyer, R. F. (1970), "The Performance of Chemical Propellants," Chapter II. Technivision, London.

Gordon, S., and McBride, B. J. (1971). NASA SP-273.

Huff, V. N., and Morell, V. E. (1950). NACA TN 1113.

JANNAF Thermochemical Data, Dow Chemical Co., Midland, Michigan.

National Bureau of Standards Circ. C461 (1947).

Penner, S. S. (1957). "Chemistry Problems in Jet Propulsion," Chapter VI. Pergamon, Oxford.

Chapter Two

Chemical Kinetics

Flames will propagate through only those chemical mixtures which are capable of reacting sufficiently fast to be considered explosive in character. Indeed the expression "explosive" is meant to specify very rapid reaction. From the standpoint of combustion, the interest in chemical kinetic phenomena is directed, generally, towards considering the conditions under which chemical systems would undergo explosive reaction. More recently, however, great interest has developed in the rates and mechanisms of steady (nonexplosive) chemical reactions because most of the complex pollutants form in zones of steady, usually lower temperature, reactions in the combustion process.

Certain essential features of chemical kinetics occur frequently in combustion phenomena. These features will be reviewed here. For a more detailed understanding of any of these aspects, and a thorough coverage of the subject, one should refer to any of the books on chemical kinetics, such as that by Benson (1960).

A. THE RATES OF REACTIONS AND THEIR TEMPERATURE DEPENDENCY

All chemical reactions, whether of the hydrolysis, acid–base or combustion type, take place at a definite rate, depending on the conditions of the system. The most important conditions are the concentration of the reactants, the temperature, radiative effects, and the presence of a catalyst or inhibitor. The rate of the reaction may be expressed in terms of the concentration of any of the substances reacting or any product of the reaction; i.e., the rate may be expressed as the rate of decrease of the concentration of a reactant or the rate of increase of a product of reaction.

Penner (1955) represents the stoichiometric relation describing a one-step chemical reaction of arbitrary complexity by the equation

$$\sum_{j=1}^{n} \nu_j' M_j \longrightarrow \sum_{j=1}^{n} \nu_j'' M_j'' \tag{1}$$

where ν_j' is the stoichiometric coefficient of the reactants, ν_j'' the stoichiometric coefficients of the products, M the arbitrary specification of all chemical species, and n the total number of compounds involved. If a species represented by M_j does not occur as a reactant or product, its ν_j equals zero. Consider, for example, the recombination of H atoms in the presence of H atoms; i.e., the reaction

$$\mathrm{H + H + H \longrightarrow H_2 + H}$$

$$n = 2, \quad M_1 = \mathrm{H}, \quad M_2 = \mathrm{H_2}; \qquad \nu_1' = 3, \quad \nu_1'' = 1, \quad \nu_2' = 0, \quad \nu_2'' = 1$$

The reason for following this complex notation will become apparent shortly.

The law of mass action, which is confirmed experimentally, states that the disappearance of a chemical species is proportional to the product of the concentrations of the reacting chemical species, each concentration raised to a power equal to the corresponding stoichiometric coefficient; i.e.,

$$RR = k \prod_{j=1}^{m} (M_j)^{\nu_j'} \tag{2}$$

where k is the proportionality constant called the specific reaction rate constant. The $\sum \nu_j'$ is also given the symbol n and is the overall order of the reaction. In an actual reacting system, the rate of change of the concentration of a given species i is given by

$$d(M_i)/dt = [\nu_i'' - \nu_i']RR = [\nu_i'' - \nu_i']k \prod_{j=1}^{m} (M_j)^{\nu_j'} \tag{3}$$

since ν_i'' moles of M_i are formed for every ν_i' moles of M_i consumed. For the previous example, $d(H)/dt = 2k(H)^3$. The use of this complex scheme prevents error in sign and eliminates confusion when stoichiometric coefficients are different from 1.

Most chemical reactions take place due to the collisions of two reactants which may have the capability to react. Thus most simple chemical reactions are second order. The literature is known to report that some reactions appear to be first order. Most of these reactions fall in the class of decomposition processes. According to Lindemann's theory (1922) of first-order processes, first-order reactions occur as a result of a two-step process. This point will be discussed later.

An arbitrary second-order reaction may be written as

$$\begin{array}{ccc} \mathrm{A} + \mathrm{B} & \longrightarrow & \mathrm{C} + \mathrm{D} \\ [\mathrm{O} + \mathrm{N_2} & \longrightarrow & \mathrm{NO} + \mathrm{N}] \end{array} \tag{4}$$

For this reaction the rate expression takes the form

$$RR = -d(\mathrm{A})/dt = +k(\mathrm{A})(\mathrm{B}) = +d(\mathrm{C})/dt \tag{5}$$

Throughout this book the convention is used such that parentheses around a chemical symbol means the concentration of that species in moles or mass per cubic centimeter. Sometimes C with the chemical symbol as a subscript will be used to specify concentration. Specifying the reaction in this manner (Eq. (4) or (5)) does not mean every collision of the reactants A and B would lead to products or cause the disappearance of an amount of A and B. Arrhenius (1889) put forth a simple theory that accounts for this fact and gives the temperature dependence of k. Arrhenius stated that only those molecules which possess energy greater than a certain amount E_A will react. Molecules acquire the additional energy necessary from collisions which produce energy in excess of E_A. These high energy active molecules lead to products. Arrhenius' postulate may be written as

$$RR = Z_{AB} \exp\{-E_A/RT\} \tag{6}$$

where Z_{AB} is the collision frequency and $\exp\{-E_A/RT\}$ is the Boltzmann factor. The Boltzmann factor as derived in kinetic theory specifies the fraction of all collisions that have an energy greater than E_A. From simple kinetic theory

$$Z_{AB} = (\mathrm{A})(\mathrm{B})\sigma_{AB}[8\pi k_B T/\mu]^{1/2} \tag{7}$$

where σ_{AB} is the collision diameter, k_B is the Boltzmann constant and μ is the reduced mass. Z_{AB} may be written in the form

$$Z_{AB} = Z'_{AB}(\mathrm{A})(\mathrm{B})$$

Thus the Arrhenius form for the rate is

$$RR = Z'_{AB}(\mathrm{A})(\mathrm{B}) \exp\{-E_A/RT\}$$

When compared to the reaction rate written from the law of mass action, the result is found that

$$k = Z'_{AB} \exp\{-E_A/RT\} = Z''_{AB} T^{1/2} \exp\{-E_A/RT\} \tag{8}$$

Thus the important conclusion is that the specific reaction rate constant is dependent on the temperature alone and is independent of the concentration. Actually, when complex molecules are reacting, not every collision has the proper steric orientation for the specified reaction to take place. Thus k can be written as

$$k = Z''_{AB} T^{1/2} \exp\{-E_A/RT\} \mathscr{P} \tag{9}$$

where $\mathscr{P}$ is an experimentally determined steric factor, which can be a very small number. Most generally the Arrhenius form of the reaction rate is written as

$$k = \text{const} \exp\{-E_A/RT\} = \text{A} \exp\{-E_A/RT\} \tag{10}$$

where the constant A takes into account the collision terms, the mild temperature dependence, and the steric factor. This form of expression holds well for most reactions, shows an increase of k with T, and permits convenient straight line correlation of data on a ln k versus $(1/T)$ plot. There are two classes of reaction for which this expression does not hold. For low activation energy free radical reactions, the temperature dependence in the preexponential takes on more importance. In this case the approach known as the absolute theory of reaction rate seems to give better correlation of reaction data with temperature.

In this theory (see Benson, 1960), the reactants are in equilibrium with an activated complex which forms. One of the vibrational modes in the complex is considered loose and permits the complex to dissociate to products. When the equilibrium constant is written in terms of the partition functions and the frequency of the loose vibration allowed to approach zero, a rate constant expression is obtained. This expression takes the form

$$k = (k_B T/h)[(Q_{T-1})^{\#}/(Q_T)_A(Q_T)_B] \exp\{-E_A/RT\} \tag{11}$$

where Q_T is the total partition function, h is Planck's constant, and E_A the small activation energy to form the complex. $(Q_{T-1})^{\#}$ is the partition function of the activated complex evaluated for all frequencies except the loose one. Consider the reaction between hydroxyl radical and CO (Dryer *et al.*, 1971).

$$\text{HO} + \text{CO} \rightleftharpoons (\text{HOCO})^{\#} \longrightarrow \text{H} + \text{OCO}$$

In this case $(\text{HOCO})^{\#}$ is the activated complex and the O—H bond is the loose vibration in the complex. Generally when radical reactions do not plot as straight lines on ln k as $(1/T)$ (Arrhenius) plots, it is best to apply this so-called absolute theory.

Radical recombination is another class of reactions in which the Arrhenius expression will not hold. When simple radicals recombine to form a single product, the energy liberated in the process is sufficiently great to cause the product to decompose into the original radicals. Energy must be removed from the product upon its formation in order to stabilize it. A third body is necessary to remove this energy. If one follows the approach of Landau and Teller (1936), who in dealing with vibrational relaxation developed an expression by averaging a transition probability based on the relative molecular velocity over the Maxwellian distribution, the following expression for the rate constant is obtained [see Vincenti and Kruger (1965)]

$$k \sim \exp(-T^{-1/3}) \tag{12}$$

However, as the temperature increases, the probability of redissociation from the higher vibrational states increases and the total rate of recombination actually decreases with a dependency approaching T^{-1} at high temperatures. In dealing with the recombination of radicals in nozzle flow, this mild temperature dependency should be kept in mind. Recall the example of H atom recombination given earlier. If the third bodies in the system are written as M, then the equation takes the form

$$\mathrm{H} + \mathrm{H} + \mathrm{M} \longrightarrow \mathrm{H_2} + \mathrm{M}$$

The rate of formation of H_2 is

$$d(\mathrm{H_2})/dt = k(\mathrm{H})^2(\mathrm{M})$$

Thus in expanding dissociated gases through a nozzle, the velocity increases, and the temperature and pressure decrease. Thus the rate constant for this process increases but only slightly. The pressure affects the concentrations and thus enters the rate as a cubed term. In all, then, the rate of recombination in the high velocity regions decreases due to the pressure term. The point to be made is that third-body recombination reactions are mostly pressure sensitive, most generally are favored at higher pressure, and rarely occur at very low pressures.

B. SIMULTANEOUS INTERDEPENDENT AND CHAIN REACTIONS

In complex reacting systems, such as those which exist in combustion processes, a simple one-step rate expression will not suffice. More generally one finds simultaneous, interdependent reactions or chain reactions.

The most frequently occurring simultaneous, interdependent reaction mechanism is the case in which the product, as its concentration is increased, begins to dissociate into the reactants. The classical example is the hydrogen–iodine reaction:

$$\mathrm{H_2} + \mathrm{I_2} \underset{k_b}{\overset{k_f}{\rightleftharpoons}} 2\mathrm{HI} \tag{13}$$

The rate of formation of HI is then affected by two rate constants, k_f and k_b, and is written

$$d(\mathrm{HI})/dt = 2k_f(\mathrm{H_2})(\mathrm{I_2}) - 2k_b(\mathrm{HI})^2 \tag{14}$$

At equilibrium, the rate of formation of HI is zero, and

$$2k_f(\mathrm{H_2})_{eq}(\mathrm{I_2})_{eq} - 2k_b(\mathrm{HI})_{eq}^2 = 0 \tag{15}$$

where the subscript eq designates the equilibrium concentrations. Thus

$$k_f/k_b = (\mathrm{HI})_{eq}^2/(\mathrm{H_2})_{eq}(\mathrm{I_2})_{eq} \equiv K_C \tag{16}$$

i.e., the forward and backward rate constants are related to the equilibrium constant based on concentrations (K_C). This constant is a thermodynamic quantity and is considered known. Thus the rate expression for HI becomes

$$d(\mathrm{HI})/dt = 2k_f(\mathrm{H_2})(\mathrm{I_2}) - 2k_b(\mathrm{HI})^2 = 2k_f(\mathrm{H_2})(\mathrm{I_2}) - 2(k_f/K_C)(\mathrm{HI})^2 \tag{17}$$

which shows there is only one independent rate constant in the problem.

In most instances, two reacting molecules do not react directly as H_2 and I_2 do, but react by one dissociating first to form radicals, which then initiate a chain of steps. Interestingly, this procedure occurs in the reaction of H_2 with another halogen Br_2. Experimentally, Bodenstein (1913) found the rate of formation of HBr to obey the expression

$$\frac{d(\mathrm{HBr})}{dt} = \frac{k'_{\mathrm{exp}}(\mathrm{H_2})(\mathrm{Br_2})^{1/2}}{1 + k''_{\mathrm{exp}}[(\mathrm{HBr})/(\mathrm{Br_2})]} \tag{18}$$

This expression shows that HBr is inhibiting to its own formation.

Bodenstein explained this result by suggesting that the H_2–Br_2 reaction was chain in character and initiated by a radical ($Br\cdot$). He proposed the following steps:

$Br_2 \xrightarrow{k_1} 2Br\cdot$		chain initiating step
$Br\cdot + H_2 \xrightarrow{k_2} HBr + H\cdot$		
$H\cdot + Br_2 \xrightarrow{k_3} HBr + Br\cdot$		chain carrying step
$H\cdot + HBr \xrightarrow{k_4} H_2 + Br\cdot$		
$2Br\cdot \xrightarrow{k_5} Br_2$		chain breaking

The bond energy in Br_2 is 46 kcal/mole and in H_2 it is 104 kcal/mole. Consequently, over all but the very highest temperatures, the Br_2 dissociation will be the initiating step. Actually the third-body symbol (M) should be added to both sides of reactions 1 and 5; however, Bodenstein did not write the reactions in this manner, and because of other assumptions to be made in determining the HBr formation rate, third-body considerations do not enter. Reaction 4 is the backward step of reaction 2. It is more convenient in analyzing complex problems to write the steps individually in this manner. The inverse of reaction 3 proceeds very slowly, is therefore not important in the system, and is omitted.

The different type of chain steps are designated in the reaction system as written above. There exists another chain step, which is perhaps the most important of the various chain types in that this step is necessary to achieve a nonthermal explosion. It is one in which two radicals are created for each radical consumed. It is called chain branching, and two typical chain branching steps that occur in the H_2–O_2 reaction are

$$H\cdot + O_2 \longrightarrow \cdot OH + \cdot O\cdot$$

$$\cdot O\cdot + H_2 \longrightarrow \cdot OH + H\cdot$$

Branching will usually occur when a monoradical ($H\cdot$) formed by breaking a single bond reacts with a species containing a double bond (O_2) or when a biradical ($\cdot O\cdot$) formed by breaking a double bond reacts with a saturated molecule (H_2 or RH). A dot written next to a chemical system is the convention for designating a radical. Since it is obvious when a radical exists, the convention is not used consistently throughout this book, unless clarity is needed.

From the five chain steps written above, the expression for the HBr formation rate is

$$d(\mathrm{HBr})/dt = k_2(\mathrm{Br})(\mathrm{H_2}) + k_3(\mathrm{H})(\mathrm{Br_2}) - k_4(\mathrm{H})(\mathrm{Br}) \tag{19}$$

In actual experimental systems, it is very difficult to measure the concentrations of radicals and it is desirable to have them expressed in terms of other known or measurable quantities. It is possible to achieve this objective by making the so-called steady state assumption for the reaction's radical intermediates. The assumption is that the radicals react so rapidly once they are formed that their concentrations do not continue to rise but reach steady state concentration. Thus the rate of formation equations for the radicals are written and then set equal to zero. For the H_2–Br_2 system, then,

$$d(\mathrm{H})/dt = k_2(\mathrm{Br})(\mathrm{H_2}) - k_3(\mathrm{H})(\mathrm{Br_2}) - k_4(\mathrm{H})(\mathrm{HBr}) = 0 \tag{20}$$

$$d(\mathrm{Br})/dt = 2k_1(\mathrm{Br_2}) - k_2(\mathrm{Br})(\mathrm{H_2}) + k_3(\mathrm{H})(\mathrm{Br_2}) + k_4(\mathrm{H})(\mathrm{HBr}) - 2k_5(\mathrm{Br}) = 0 \tag{21}$$

To reiterate, setting the above two expressions equal to zero does not mean equilibrium conditions exist. In this case it means a steady state situation has been reached, and the steady state concentrations of Br and H become

$$(\mathrm{Br}) = (k_1/k_5)^{1/2}(\mathrm{Br}_2)^{1/2} \tag{22}$$

$$(\mathrm{H}) = \frac{k_2(k_1/k_5)(\mathrm{H}_2)(\mathrm{Br}_2)^{1/2}}{k_3(\mathrm{Br}_2) + k_4(\mathrm{HBr})} \tag{23}$$

Substituting these values in the rate expression for HBr (Eq. (19)), one obtains

$$\frac{d(\mathrm{HBr})}{dt} = \frac{2k_2(k_1/k_5)^{1/2}(\mathrm{H}_2)(\mathrm{Br}_2)^{1/2}}{1 + (k_4/k_3)[(\mathrm{HBr})/(\mathrm{Br}_2)]} \tag{24}$$

which is the exact form found experimentally (Eq. (18)). Thus

$$k'_{\mathrm{exp}} = 2k_2(k_1/k_5)^{1/2}, \qquad k''_{\mathrm{exp}} = k_4/k_3$$

By the steady state assumption, the rate constant for the intermediate steps can be determined. Since k'_{exp} and k''_{exp} can be determined experimentally, the individual rate constants follow directly. Since (k_1/k_5) is the equilibrium constant for Br_2 dissociation and known, k_2 is found from k'_{exp}. k_4 is found from k_2 and the equilibrium constant that represents reactions 2 and 4.

C. PSEUDO-FIRST-ORDER REACTIONS

As mentioned earlier, practically all reactions take place by bimolecular collisions; however, certain reactions exhibit first-order kinetics. This possible anomaly has been explained by Lindemann (1922), who proposed that a first-order process occurred as a result of a two-step reaction such as

$$\mathrm{A}_2 + \mathrm{A}_2 \underset{k_b}{\overset{k_f}{\rightleftharpoons}} \mathrm{A}_2^* + \mathrm{A}_2 \qquad \text{(fast)} \tag{25}$$

$$\mathrm{A}_2^* \xrightarrow{k_f'} \text{products} \qquad \text{(slow)} \tag{26}$$

The above process remains first order as long as the step designated "fast" stays fast enough to maintain an equilibrium concentration of A_2^*, the activated higher energy complex that dissociates. Since the frequency of binary collisions decreases with pressure, one would expect the first step to slow down with pressure and the overall order of the reaction to become second at low pressures. Indeed, this change in order does occur for all collisional reaction mechanisms that appear first order at high pressures. A steady state analysis on A_2^* for the Lindemann process steps above gives

$$d(\mathrm{A}_2)/dt = -k_f k_f'(\mathrm{A}_2{}^2)/[k_b(\mathrm{A}_2) + k_f'] \tag{27}$$

which shows the order change with pressure. At high pressure $k_b(A_2) \gg k_f'$; at low pressure, vice versa.

Sometimes reaction measurements appear to indicate a first-order mechanism but are actually pseudo-first order. Consider the arbitrary reaction process

$$A + B \longrightarrow D$$

where (B) $\gg$ (A). Then the rate expression would be

$$d(A)/dt = -d(D)/dt = -k(A)(B)$$

Since (B) $\gg$ (A), the concentration of (B) does not change appreciably, and $k(B)$ would appear as a constant. If $k(B)$ were written as k', then

$$d(A)/dt = -d(D)/dt = -k'(A)$$

where $k' = k(B)$. This expression appears first order, but one should notice that k', since it contains a concentration (B), is pressure dependent. This pseudo-first order concept arises in many practical combustion systems that are very fuel lean.

D. PRESSURE EFFECT IN FRACTIONAL CONVERSION

In combustion problems, the rate of energy conversion or utilization is of greatest interest. Thus it is more convenient to deal with the fractional change of a particular substance rather than the absolute concentration. If C is used to denote the concentrations in a chemical reacting system of arbitrary order n, then the rate expression is

$$dC/dt = -kC^n$$

Since C is a concentration, it may be written in terms of the total density ρ and the mole fraction ε,

$$C = \rho\varepsilon$$

It follows then for a constant temperature system

$$\rho\, d\varepsilon/dt = -k(\rho\varepsilon)^n, \qquad d\varepsilon/dt = -k\varepsilon^n\rho^{n-1}$$

For a constant temperature system, $\rho \sim P$ and

$$d\varepsilon/dt \sim P^{n-1}$$

The fractional change is proportional to the pressure to the order minus 1.

REFERENCES

Arrhenius, S. (1889). *Z. Phys. Chem.* **4,** 226.
Benson, S. W. (1960). "The Foundations of Chemical Kinetics." McGraw-Hill, New York.
Bodenstein, M. (1913). *Z. Phys. Chem.* **85,** 329.
Dryer, F. L., Naegeli, D. W., and Glassman, I. (1971). *Combust. Flame* **17**, 270.
Landau, L., and Teller, E. (1936). *Phys. Z. Sowjet.* **10,** 1, 34.
Lindemann, F. A. (1922). *Trans. Faraday Soc.* **17,** 598.
Penner, S. S. (1955). "Introduction to the Study of the Chemistry of Flow Processes," Chapter 1. Butterworths, London.
Vincenti, W. G., and Kruger, C. H., Jr. (1965). "Introduction to Physical Gas Dynamics," Chapter VII. Wiley, New York.

Chapter Three

Explosive and General Oxidation Characteristics of Fuels

In the previous chapters the fundamental areas of thermodynamics and chemical kinetics were reviewed. These areas provide the background for the study of the very fast reactions, termed explosions. In order for flames (deflagrations) or detonations to propagate, the reaction kinetics must be fast; i.e., the mixture must be explosive.

A. THE CRITERION FOR EXPLOSION

Consider, for example, a mixture of hydrogen and oxygen stored in a vessel in stoichiometric proportions and at a total pressure of 1 atm. The vessel is immersed in a thermal bath kept at 500°C, as shown in Fig. 1.

If the vessel shown in Fig. 1 is evacuated to a few millimeters of mercury (torr) pressure, there is an explosion. Similarly, if the system is pressurized to 2 atm pressure, there is also an explosion. These facts suggest explosive limits.

If H_2 and O_2 react to form explosive combustion, it is possible that such processes could occur in a flame, which indeed they do. A fundamental

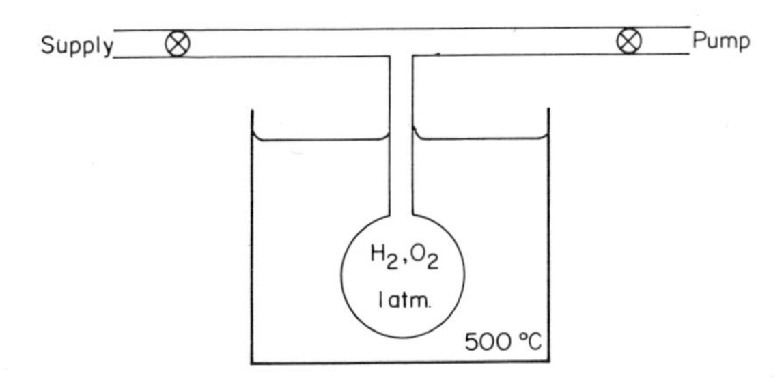

Fig. 1. Experimental apparatus used for the determination of explosion limits.

question is what governs the conditions which give explosive mixtures. To answer this question, it is well to consider again the chain reaction as it occurs in the H_2 and Br_2 reaction:

$$H_2 + Br_2 \longrightarrow 2HBr \qquad \text{(the overall reaction)}$$

$$M + Br_2 \longrightarrow 2Br + M \qquad \text{(initiation step)}$$

$$\left.\begin{array}{l} Br + H_2 \longrightarrow HBr + H \\ H + Br_2 \longrightarrow HBr + Br \\ H + HBr \longrightarrow H_2 + Br \end{array}\right\} \qquad \text{(chain cycle steps)}$$

$$M + 2Br \longrightarrow Br_2 + M \qquad \text{(chain termination)}$$

There are two means by which the reaction can be initiated—thermally or photochemically. If the H_2–Br_2 mixture is at room temperature, a photochemical experiment can be performed by using light of short wave length; i.e., high enough $h\nu$ to rupture the Br—Br bond through a transition to a higher electronic state. In an actual experiment, the light source can be made as weak as possible and the actual energy measured. Then, it is possible to estimate the number of bonds broken and measure the number of HBr molecules formed. The ratio of HBr molecules formed per Br atom created is called the photoyield. It is found in the room temperature experiment that

$$HBr/Br \sim 0.01 \ll 1$$

and, of course, no explosive characteristic is observed because the reaction

$$Br + H_2 \longrightarrow HBr + H$$

is quite endothermic and thus slow. Therefore the chain effect is overtaken by the recombination reaction

$$M + 2Br \longrightarrow Br_2 + M$$

Thus there are competitive reactions which appear to determine the overall character of the reacting system.

For the system H_2–Cl_2, the photoyield is of the order 10^4–10^7. In this case, the chain step is much faster in that the reaction

$$Cl + H_2 \longrightarrow HCl + H$$

has an activation energy of only 6 kcal/mole compared to 18 kcal/mole for the corresponding bromine reaction. The fact that in the iodine reaction the corresponding step has an activation energy of 33 kcal/mole gives credence to the fact that the iodine reaction does not proceed through a chain mechanism, whether initiated thermally or by photolysis.

From the above discussion, it is obvious that only the H_2–Cl_2 reaction can be exploded photochemically, i.e., at low temperatures. The H_2–Br_2 and H_2–I_2 systems can only support thermal (higher temperature) explosions.

Recall in the discussion of kinetic processes it was emphasized that the H_2–O_2 reaction contains an important characteristic chain branching step, namely,

$$H + O_2 \longrightarrow OH + O$$

which leads to a further chain branching system,

$$O + H_2 \longrightarrow OH + H$$

$$OH + H_2 \longrightarrow H_2O + H$$

The first two of the above three steps are branching, in that two radicals are formed for each one consumed. Since all three steps are necessary in the chain system, the multiplication factor, usually designated α, is seen to be greater than 1 but less than 2. The first of the above three reactions is strongly endothermic and thus will not proceed rapidly at low temperatures. So at low temperature an H atom can survive many collisions and can find its way to a surface to be destroyed. This result explains why there is steady reaction in some H_2–O_2 systems when H radicals are introduced. Explosion occurs only at the higher temperatures where the first step proceeds more rapidly.

It is interesting to consider the effect of the multiplication factor. In a particular straight chain reaction, assume there are 10^8 collisions/sec, 1 chain particle/cm^3, and 10^{19} molecules/cm^3. Thus the molecules will be consumed in 10^{11} sec or approximately 30 years.

In a particular branched chain reaction, the same basic conditions as before are assumed. However, the multiplication factor is taken as 2. Thus

$$2^N = 10^{19}, \qquad N = 62$$

All the molecules are consumed in 62 generations, and the time for completion is 62×10^{-8} sec or approximately 10^{-6} sec. For $\alpha = 1.01$, the time is only 10^{-4} sec, consequently one may conclude that as long as $\alpha > 1$, the reaction proceeds rapidly.

A general branched chain reaction system may be written as

$$\mathrm{M} \xrightarrow{k_1} \mathrm{R} \qquad \text{initiation}$$

$$\mathrm{R} + \mathrm{M} \xrightarrow{k_2} \alpha\mathrm{R} + \mathrm{M}^* \qquad \text{chain branching, } \alpha > 1$$

$$\mathrm{R} + \mathrm{M} \xrightarrow{k_3} \mathrm{P} \qquad \text{product formation, removes radical}$$

$$\left.\begin{array}{l} \mathrm{R} \xrightarrow[\text{wall}]{k_4} \text{destruction} \\ \mathrm{R} \xrightarrow[\text{gas}]{k_5} \text{destruction} \end{array}\right\} \text{chain termination}$$

where M is a molecule, R a radical, and P a product. Now the question to be considered is what value of α is necessary for the system to be explosive. A simple steady state analysis is used. The explosive condition is determined by the rate of formation of the product,

$$d(\mathrm{P})/dt = k_3(\mathrm{R})(\mathrm{M}) \tag{1}$$

The steady state condition for the radicals is

$$d(\mathrm{R})/dt = 0 = k_1(\mathrm{M}) + k_2(\alpha - 1)(\mathrm{R})(\mathrm{M}) - k_3(\mathrm{R})(\mathrm{M}) - k_4(\mathrm{R}) - k_5(\mathrm{R}) \tag{2}$$

Solving for (R) and substituting into the product rate equation, one obtains

$$d(\mathrm{P})/dt = k_1 k_3(\mathrm{M})^2/\{k_3(\mathrm{M}) + k_4 + k_5 - k_2(\alpha - 1)(\mathrm{M})\} \tag{3}$$

The rate of formation of product becomes infinite, or the system explosive, when the denominator equals zero. Thus, a critical value of α can be specified as

$$\alpha_{\mathrm{crit}} = 1 + \frac{k_3(\mathrm{M}) + k_4 + k_5}{k_2(\mathrm{M})} = \left(1 + \frac{k_3}{k_2}\right) + \frac{k_4 + k_5}{k_2(\mathrm{M})} \tag{4}$$

such that for $\alpha_{\mathrm{react}} > \alpha_{\mathrm{crit}}$, the system is explosive; for $\alpha_{\mathrm{react}} < \alpha_{\mathrm{crit}}$, the products form by slow reaction. For most purposes, (M) is proportional to the total pressure, and one readily can understand the pressure effect observed for the H_2–O_2 system discussed at the beginning of this section.

B. EXPLOSION LIMITS AND OXIDATION CHARACTERISTICS OF HYDROGEN

Much has been learned about the combustion mechanisms of hydrogen by the study of its explosive limits. There have been extensive treatises written on the subject of the H_2–O_2 reaction. In particular, much attention has been given to the effect of the wall on radical destruction. Such effects are not important in many of the combustion reactions which are of fundamental interest in this volume. Thus this aspect of the problem will not be emphasized. Although there is no fundamental agreement as to all the steps in the reaction, there is general agreement as to the basic mechanism. It is interesting that the basic mechanism can be inferred from the characteristic explosion limits of the reaction.

It is now important to stress some points in order to eliminate possible confusion with previously held concepts and certain items to be discussed later. The explosive limits are not flammability limits. Explosion limits are the pressure–temperature boundaries for a specific mixture ratio of fuel and oxidizer and separate the regions of slow and fast reaction. For a specified temperature and pressure, flammability limits specify the lean and rich fuel mixture ratio beyond which no flame will propagate. One must have fast reaction for a flame to propagate. A stoichiometric mixture of H_2 and O_2 at standard conditions will support a flame because an ignition source initially brings a local mixture into the explosive regime and the established flame, through diffusive mechanisms, heats fresh mixture to high enough temperatures to be explosive. Thus, in the early parts of the flame, the mixture follows steady reaction and in the latter parts, explosive reactions. This point is significant, particularly in hydrocarbon combustion, because it is in the low temperature regime where particular compounds that lead to pollution problems are formed.

Given in Figure 2 are the explosion limits of a stoichiometric mixture of H_2 and O_2. Explosion limits can be found for many different mixture ratios. The point $\times$ on Fig. 2 marks the conditions (500°C–1 atm) described at the very beginning of this chapter in Fig. 1. It now becomes obvious that increasing or decreasing the pressure at constant temperature can cause an explosion.

Certain general characteristics of this curve can be stated. The third limit portion of the curve is as one would expect from simple density considerations. Any discussion of the first or lower limit will be related to wall effects and its role in chain destruction.

The expression developed for α_{crit} (Eq. (4)) applies to the lower limit only when the wall effect is considered a first-order reaction of chain destruction, since R $\xrightarrow[\text{wall}]{k_4}$ destruction was written. The three limits can be explained by reasonable hypotheses of mechanisms, although not all of the explanations are accepted by everyone and are still debated. In general, the features of the movement of the boundaries are not explained fully. The presence of radicals and metastable species makes it difficult to subject the mechanism to thorough laboratory analysis. The advent of refined mass spectroscopic techniques has contributed significantly to this area.

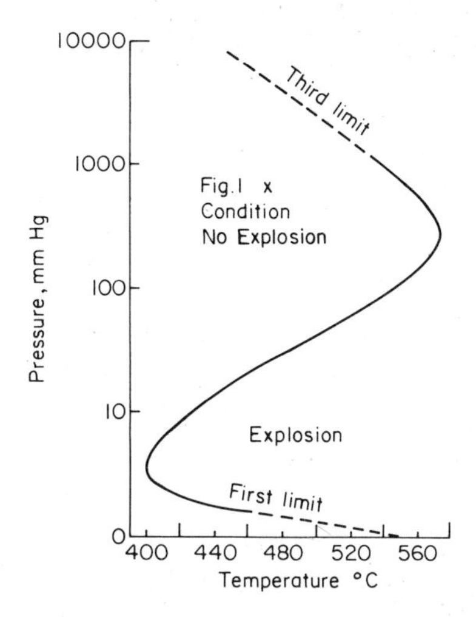

Fig. 2. Explosion limits of a stoichiometric hydrogen–oxygen mixture (after Lewis and von Elbe, 1961).

The manner in which the reaction is initiated to give the front designated by the curve in Fig. 2 suggests, as was inferred earlier, that the explosion is in itself a branched chain phenomenon. Thus possible branched chain mechanisms to explain the limits must be considered.

Basically, thermal rather than photolysis mechanisms are considered. The dissociation energy of hydrogen is less than oxygen, so that the initiation can be related to the hydrogen dissociation. Only a few radicals are required to initiate the explosion in the region of temperature of interest, i.e., above 400°C.

If the hydrogen dissociation is the chain's initiating step, then it proceeds by the reaction

$$H_2 + M \longrightarrow 2H + M \tag{5}$$

which requires about 106 kcal/mole. However, an abundance of radicals can come from the step

$$M + H_2 + O_2 \longrightarrow \underset{\substack{\downarrow \\ 2OH}}{H_2O_2} + M^* \qquad +51 \quad \text{kcal/mole} \tag{6}$$

This reaction requires only 51 kcal/mole but is trimolecular. Nevertheless, it is apparently the primary reaction for chain initiation at the lower temperatures, and there is some plausibility in using OH formation as the initial step in the chain formation. The previous chain propagating steps should then be written in the order

$$OH + H_2 \longrightarrow H_2O + H \qquad -15 \text{ kcal/mole} \tag{7}$$

$$H + O_2 \longrightarrow OH + O \qquad +16 \text{ kcal/mole} \tag{8}$$

$$O + H_2 \longrightarrow OH + H \qquad +2 \text{ kcal/mole} \tag{9}$$

although either OH or H can establish the same steps. The reverse reactions of Eqs. (8) and (9) can be neglected because they involve a binary collision of two active radical species, which are present in very small concentrations. Since all constituents are found in flames, reactions (7)–(9) are undoubtedly the proper chain.

The most important chain destruction mechanism is

$$H \longrightarrow \text{wall destruction} \tag{10}$$

or

$$OH \longrightarrow \text{wall destruction} \tag{11}$$

either of which explains the lower limit of explosion. This result is apparent since wall collisions become relatively more predominate at lower pressures than molecular collisions do. The fact that the limit is found experimentally to be a function of diameter is further evidence of this type of step.

The second limit must be explained by gas phase production and gas phase destruction of radicals. It is found to be independent of vessel diameter. To have this limit, the most effective chain-branching reaction (8) must be destroyed and thus there must be a third-order reaction to take over the branching second-order reaction (8). Further, two of the molecules in the third-order reaction must be in abundance to be effective. The following appears to satisfy all three prerequisites:

$$H + O_2 + M \longrightarrow HO_2 + M \qquad -48 \text{ kcal/mole} \tag{12}$$

M is the usual third body which takes away the energy and stabilizes the combination of H and O_2. At higher pressures, it certainly is possible to get proportionally more of this trimolecular reaction than the binary reaction

$$H + O_2 \longrightarrow OH + O$$

The free radical HO_2 is thought to be relatively unreactive so that it is able to diffuse to the wall and it becomes a vehicle for destruction of free radicals. Its existence has been established by mass spectrometry, particularly in electric discharges. Molecules, such as HO_2, are referred to as metastable since they are not considered stable species but are long-lived. The destruction reactions are first order and could be both of the following

$$HO_2 \xrightarrow{\text{wall}} \tfrac{1}{2}H_2 + O_2 \tag{13}$$

$$\xrightarrow{\text{wall}} H_2O + \tfrac{1}{2}O_2$$

The upper (third) explosive limit is due to a reaction which overtakes the stability of the HO_2, most possibly the system

$$HO_2 + H_2 \longrightarrow H_2O_2 + H \tag{14}$$

which regenerates a large number of active particles. In essence, what is inferred is that if the pressure is increased and all other factors are the same, the probability of reaction (14) increases relative to the probability of diffusion of HO_2 to a wall for destruction.

Water vapor tends to inhibit explosion due to the effect of reaction (12) on the 2nd limit

$$H + O_2 + M \longrightarrow HO_2 + M$$

in that water is a most effective third body probably due to some resonance energy exchange.

At temperatures well above 600°C, it is not likely that the HO_2 molecule can be stabilized due to the energy of the collision partners and thus the mixture is explosive at all pressures of interest. Thus in flames the chain is not necessarily terminated, certainly not by wall effects. But the radicals, being nonstable species, will recombine.

$$H + OH + M \longrightarrow H_2O + M \qquad -119\ \text{kcal/mole}$$

$$H + H + M \longrightarrow H_2 + M \qquad -104\ \text{kcal/mole}$$

$$O + O + M \longrightarrow O_2 + M \qquad -120\ \text{kcal/mole}$$

Therefore the slow combustion reactions of hydrogen which lead to explosion include initiating steps primarily proceeding through hydrogen peroxide to form hydroxyl radical or through the H radical formed through the dissociation of H_2. Reactions with the hydroperoxyl radical play a significant role. At higher temperature, the chain is supported more directly by the hydrogen dissociation. Since the hydroperoxyl radical cannot be stabilized as readily by high energy third bodies, its presence at the higher temperature is not likely. Thus at higher temperatures there are no limits, and the explosive regime exists everywhere. At still higher temperatures, where the concentration of the radicals increases, the recombination reactions begin to play a role. In most flame phenomena, whether hydrogen or a hydrocarbon is the fuel, it is the regime at the high temperature end that is important. This end is at the adiabatic flame temperature for deflagrations and the Chapman–Jouguet temperature for detonations. Thus in flame phenomena, the explosion regime is the one of importance. The steady, higher temperature regime is important with respect to induction times with the possible formation of intermediate compounds (particularly in hydrocarbon combustion) in lower temperature boundary layers and quench areas. It is these lower temperature areas, again particularly in hydrocarbon combustion, which contribute substantially to the amount of unburned hydrocarbons and oxygenates.

C. EXPLOSION LIMITS AND OXIDATION CHARACTERISTICS OF CARBON MONOXIDE

There is much less agreement about the elementary oxidation mechanisms of dry carbon monoxide than there is with hydrogen. In fact, there is recent evidence that the mechanism which was gaining favor is possibly not correct. It is very important to note that the presence of any hydrogen containing material can completely alter the picture and, in fact, there is agreement in the oxidation of "wet" carbon monoxide. Only 20 ppm of hydrogen can change the complete mechanism of carbon monoxide; thus in most practical systems, carbon monoxide will proceed through this so-called "wet" route.

It is informative, however, to try to elucidate the mechanisms for dry CO oxidation. Again, the approach is to consider the explosive limits of a stoichiometric, dry $CO-O_2$ mixture. The explosive limits to be given and their reproducibility are not well defined, principally due to the fact that the extent of dryness in each experiment may not be the same. Thus, typical results for explosion limits for "dry" CO would be as depicted in Fig. 3.

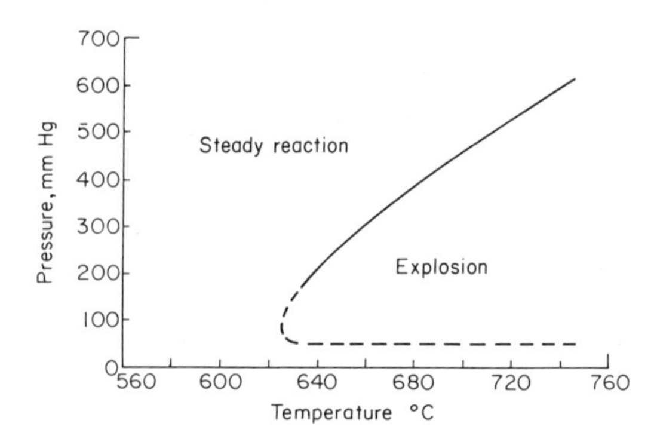

Fig. 3. The explosive peninsula of a carbon monoxide–oxygen mixture (after Lewis and von Elbe, 1961).

It is therefore seen that the low pressure ignition of $CO-O_2$ is characterized by an explosion peninsula, just as in the case of H_2-O_2. Outside this peninsula a pale-blue glow is often observed and its limits determined as well. A third limit has not been defined, and if it exists, it lies well above 1 atm.

Again, certain general characteristics of the curve may be stated. The lower limit meets all the arguments of wall killing of a chain propagating species. The effect of vessel diameter and surface has been established by experiment. There need be no discussion of the third limit since it is not well defined.

It is generally agreed that the most likely chain initiating step in the dry combustion is the reaction

$$CO + O_2 \longrightarrow CO_2 + O \qquad -9 \text{ kcal/mole}$$

which is exothermic and should proceed readily. The further steps in the oxidation of CO undoubtedly involve O atoms but the exact nature of the further reactions is in dispute. Lewis and von Elbe (1961) suggested that chain branching occurred through ozone in the following way

$$O + O_2 + M \longrightarrow O_3 + M$$

This reaction is slow but could build up in supply. Ozone is the metastable species in the process (like HO_2 in H_2–O_2 explosions), could give chain-branching, and help explain the explosion limits. The branching would come about from the step

$$O_3 + CO \longrightarrow CO_2 + 2O$$

Ozone destruction at the wall to form oxygen would explain the lower limit. Lewis and von Elbe (1961) explain the upper limit by the following third-order reaction

$$O_3 + CO + M \longrightarrow CO_2 + O_2 + M$$

However, O_3 does not appear to react with CO below 523°K. Since the CO is apparently oxidized by the oxygen atoms formed by the decomposition of ozone (the reverse of the formation of ozone), the reaction must have a high activation energy (>30 kcal/mole). This oxidation of CO by O atoms is very rapid in the higher temperature range.

Analysis of the glow and emission spectra of the CO–O_2 reaction suggests that excited carbon dioxide molecules could be present. If it is argued that the O atoms cannot react with oxygen (to form ozone), then they must react with the carbon monoxide.

A suggestion of Semenov was further developed by Gordon and Knipe (1955) to give the following alternate scheme for chain branching

$$CO + O \longrightarrow CO_2^*$$

$$CO_2^* + O_2 \longrightarrow CO_2 + 2O$$

This process is exothermic and might be expected to occur. Gordon and

Knipe (1955) counter the objection that CO_2^* is short-lived in that through system-crossing in excited states its lifetime may be sufficient to sustain the process.

In this scheme, the competitive third-body reaction to explain the upper limit becomes simply

$$CO + O + M \longrightarrow CO_2 + M$$

Most recently, however, Brokaw (1967) has disputed this mechanism, in that it does not explain the shock tube rate data. Brokaw offers the speculation that explosions in this system are thermally initiated by the nearly thermoneutral reaction

$$CO + O_2 \longrightarrow CO_2 + O$$

with subsequent large energy release through the third-body steps

$$CO + O + M \longrightarrow CO_2 + M$$

and

$$O + O + M \longrightarrow O_2 + M$$

Very early, from the analysis of ignition, flame spread, and detonation velocity data, investigators realized that small concentrations of hydrogen-containing material would appreciably catalyze the kinetics of $CO–O_2$. The H_2O catalyzed reaction proceeds in the following manner

$$CO + O_2 \longrightarrow CO_2 + O$$

$$O + H_2O \longrightarrow 2OH$$

$$OH + CO \longrightarrow CO_2 + H$$

$$H + O_2 \longrightarrow OH + O$$

If H_2 were the catalyst, then the additional steps

$$O + H_2 \longrightarrow OH + H$$

$$OH + H_2 \longrightarrow H_2O + H$$

must be included. In fact, all the steps of the $H_2–O_2$ reaction could be considered intermediaries. It is generally agreed that the important step in wet CO oxidation to form CO_2 is

$$OH + CO \longrightarrow CO_2 + H$$

This reaction is known to be quite rapid and important in later stages of hydrocarbon oxidation. Notice the analogy between this step and the step

$$OH + H_2 \longrightarrow H_2O + H$$

in the $H_2–O_2$ scheme.

As would be expected, the presence of water broadens the explosion peninsula and extends it to lower temperatures. The explanation of the limits would be the same as for the H_2–O_2 reaction since it is the OH radical which determines the CO oxidation kinetics. Since the OH is always regenerated in the chain, it is not necessary to have large quantities of the catalyst to be effective. In fact, in ignition, flame speed, and detonation results, after a very few percent of catalyst, there is no further effect.

D. EXPLOSION LIMITS AND OXIDATION CHARACTERISTICS OF HYDROCARBONS

It is interesting to note that the combustion mechanism of the most simple of all the hydrocarbons, methane, was for a long period of time the least understood. In recent years, however, there have been a great many studies of methane and its specific oxidation mechanisms are known over various ranges of temperatures. These mechanisms are now some of the best understood and the details will be discussed later in this chapter.

The higher-order hydrocarbons, particularly propane and above, burn much slower than hydrogen, but are known to form metastable molecules which were found so important in explaining the explosion limits of hydrogen and carbon monoxide. The existence of these metastable molecules makes it possible to explain qualitatively the unique explosion limits of the complex hydrocarbons and to gain some insights into what the oxidation mechanisms are likely to be.

Mixtures of hydrocarbons and oxygen react very slowly at temperatures below 200°C; as the temperature increases a variety of oxygen-containing compounds can begin to form. As the temperature is increased further, CO and H_2O begin to predominate in the products and H_2O_2 (hydrogen peroxide), CH_2O (formaldehyde), CO_2, and other compounds begin to appear. At 300–400°C a faint light often appears, which may be followed by one or more blue flames that successively traverse the reaction vessel. These light emissions are called cool flames and can be followed by an explosion. Generally, the presence of aldehydes is revealed.

In discussing the mechanisms of hydrocarbon oxidation and later the chemical reactions in photochemical smog, it becomes necessary to identify compounds which may appear complex in structure and nomenclature to those not familiar with organic chemistry. One need not have a background in organic chemistry in order to understand combustion mechanisms; one should, however, study the following section to obtain an elementary knowledge of organic nomenclature and structure.

1. Organic Nomenclature

No attempt is made to cover all the complex organic compounds which exist. The classes of organic compounds reviewed are those which occur most frequently in combustion processes and photochemical smog.

Alkyl Compounds

Paraffins (alk*anes*) (single bonds $-\overset{\vert}{\underset{\vert}{C}}-\overset{\vert}{\underset{\vert}{C}}-$)	CH_4, C_2H_6, C_3H_8, C_4H_{10}, ..., C_nH_{2n+2} meth*ane*, eth*ane*, prop*ane*, but*ane* ... straight chain iso-butane ... branched chain all are saturated (i.e., no more hydrogen can be added to any of the compounds) radicals deficient in one H atom take the names methyl, ethyl, propyl, etc.
Olefins (alk*enes*) (contain double bonds, $>C=C<$)	C_2H_4, C_3H_6, C_4H_8, ..., C_nH_{2n} eth*ene*, prop*ene*, but*ene* (ethylene, propylene, butylene) di-olefins contain two double bonds the compounds are unsaturated since C_nH_{2n} can be saturated to C_nH_{2n+2}
Cycloparaffins (cycloalkanes single bond $-\overset{\vert}{C}-\overset{\vert}{C}-$, ring closed through C)	C_nH_{2n}—no double bonds cycloprop*ane*, cyclobut*ane*, cyclopent*ane* compounds are unsaturated since ring can be broken $C_nH_{2n} + H_2 \rightarrow C_nH_{2+2}$
Acetylenes (alk*ynes*) (contain triple bonds $-C\equiv C-$)	C_2H_2, C_3H_4, C_4H_6, ..., C_nH_{2n-2} eth*yne*, prop*yne*, but*yne* (acetylene, methyl acetylene, ethyl acetylene) unsaturated compounds

Aromatics

The building block for the aromatics is the ring structured benzene, C_6H_6 which has many resonance structures and is therefore very stable.

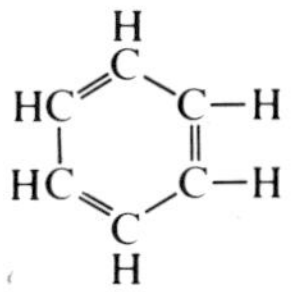

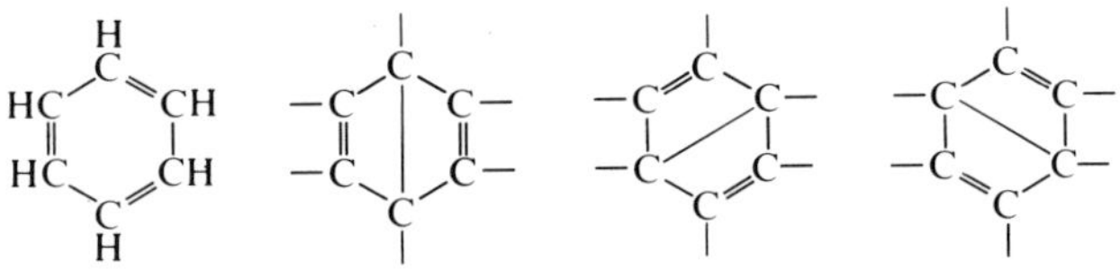

The ring structure of benzene is written in shorthand as either

or ϕ

Thus

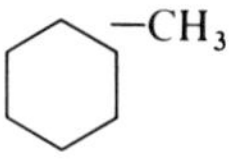

$-CH_3$ OH CH_3 CH_3

toluene 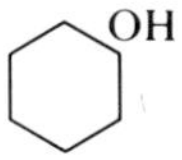 phenol (benzol) 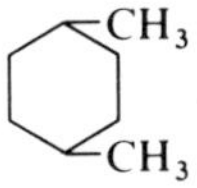 xylene

xylene being ortho, meta, or para according to whether methyl groups are separated by one, two, or three carbon atoms, respectively.

Alcohols

Those organic compounds which contain a hydroxyl group (—OH) are called alcohols and follow the simple naming procedure.

CH_3OH	C_2H_5OH
methanol (methyl alcohol)	ethanol (ethyl alcohol)

The bonding arrangement is always

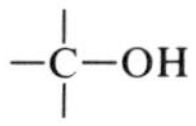

$$-\overset{|}{\underset{|}{C}}-OH$$

Aldehydes

The aldehydes contain the characteristic group (formyl radical)

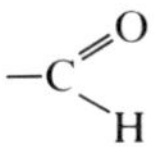

$$-C{\lower{0pt}{\overset{\nearrow O}{\searrow H}}}$$

and can be written more generally as

$$R-C\begin{matrix}{}^{\displaystyle =O}\\{}_{\displaystyle \diagdown H}\end{matrix}$$

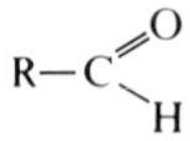

where R can be a hydrogen atom or an organic radical. Thus,

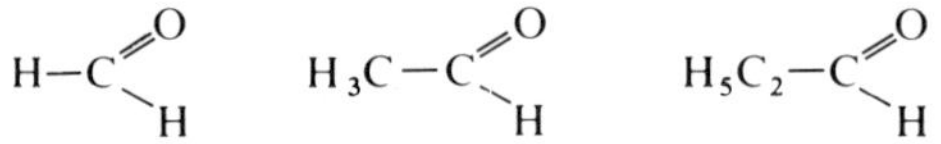

$$H-C\begin{matrix}{}^{\displaystyle =O}\\{}_{\displaystyle \diagdown H}\end{matrix} \qquad H_3C-C\begin{matrix}{}^{\displaystyle =O}\\{}_{\displaystyle \diagdown H}\end{matrix} \qquad H_5C_2-C\begin{matrix}{}^{\displaystyle =O}\\{}_{\displaystyle \diagdown H}\end{matrix}$$

formaldehyde acetaldehyde proprionaldehyde

Ketones

The ketones contain the characteristic group

$$\begin{matrix}-C-\\ \|\\ O\end{matrix}$$

and can be written more generally as

$$\begin{matrix}R'-C-R'\\ \|\\ O\end{matrix}$$

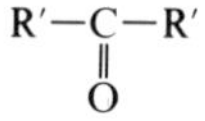

where R′ is an organic radical only. Thus,

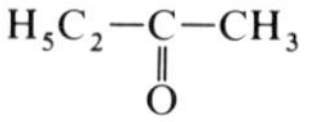

$$\begin{matrix}H_5C_2-C-CH_3\\ \|\\ O\end{matrix}$$

would be methyl ethyl ketone.

Acids

Organic acids contain the groups

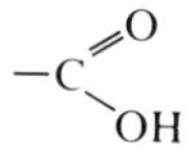

$$-C\begin{matrix}{}^{\displaystyle =O}\\{}_{\displaystyle \diagdown OH}\end{matrix}$$

or more generally

$$R-C\begin{matrix}{}^{\displaystyle =O}\\{}_{\displaystyle \diagdown OH}\end{matrix}$$

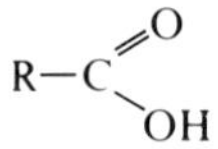

when R can be a hydrogen atom or an organic radical

$$H-C\begin{matrix}{}^{\displaystyle =O}\\{}_{\displaystyle \diagdown OH}\end{matrix} \qquad H_3C-C\begin{matrix}{}^{\displaystyle =O}\\{}_{\displaystyle \diagdown OH}\end{matrix}$$

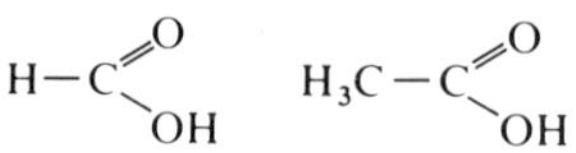

formic acid acetic acid

Organic Salts

$$\mathrm{R{-}C}\begin{smallmatrix}\nearrow\!\!\!\!=\mathrm{O}\\ \searrow\mathrm{OONO_2}\end{smallmatrix}\qquad \mathrm{HC{-}C}\begin{smallmatrix}\nearrow\!\!\!\!=\mathrm{O}\\ \searrow\mathrm{OONO_2}\end{smallmatrix}$$

peroxyacyl nitrate — peroxyacetyl nitrate PAN

Other

The ethers take the form $R^1{-}O{-}R^1$, where R^1 is an organic radical. The peroxides take the form $R^1{-}O{-}O{-}R^1$ or $R^1{-}O{-}O{-}H$, in which case the term hydroperoxide is used.

2. Explosion Limits, Cool Flames, and General Mechanisms

At temperatures of 300°–400°C and slightly higher, explosive reaction in hydrocarbon–air mixtures can take place. Thus explosion limits exist in hydrocarbon oxidation. A general representation of the explosion limits of hydrocarbons is shown in Fig. 4.

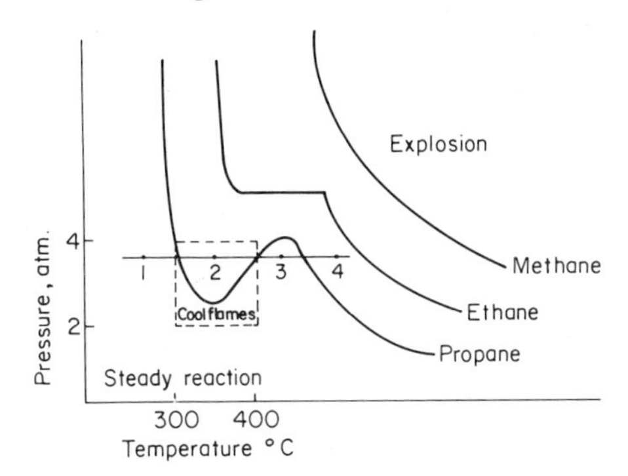

Fig. 4. General explosion limit characteristics of hydrocarbon–air mixtures. Dashed box denotes cool flame region.

One would expect the shift of curves as shown in Fig. 4, since the more complicated molecules tend to break down more readily to form radicals which promote fast reaction. The shape of the propane curve suggests that branched chain mechanisms are possible for hydrocarbons. One can conclude that the character of the propane mechanisms must be different from the H_2–O_2 reaction, when one compares the explosion curve with the H_2–O_2 pressure peninsula. The island in the propane–air curve drops and goes slightly to the left for higher order paraffins; e.g., for hexane it is located at about 1 atm. For the reaction of propane with pure oxygen, the curve drops to about 0.5 atm.

Hydrocarbons exhibit certain experimental combustion characteristics which are consistent with the explosion limit curves and which are worth reviewing:

1. They exhibit induction intervals which are followed by a very rapid reaction rate. Below 300°C these intervals are of the order of 1 min, and below 400°C, they are of the order of 1 sec or a fraction thereof.

2. Their rate of reaction is inhibited strongly by adding surface (therefore an important part of the reaction mechanism must be of the free radical type).

3. Aldehyde groups form and have an influence

$$\left(\mathrm{CHO},\ -\mathrm{C}\begin{matrix}\diagup \mathrm{H}\\ \diagdown\!\!\diagdown \mathrm{O}\end{matrix}\right)$$

They are extremely accelerating and shorten the ignition lags (formaldehyde is the strongest).

4. One finds the presence of so-called cool flames, except for methane and ethane.

5. They exhibit negative temperature coefficients of reaction rate.

6. Two stage ignition is observed and is related, perhaps, to the cool flame phenomena.

7. Explosion occurs without appreciable self-heating (branched chain explosion, without steady temperature rise) and usually occurs passing from region 1 to region 2 in Fig. 4. Explosions may occur in other regions, but the reactions are so fast one cannot tell whether they are self heating or not.

Semenov (1930) explained the long induction periods on the basis of the hypothesis that there are unstable but long-lived particles.

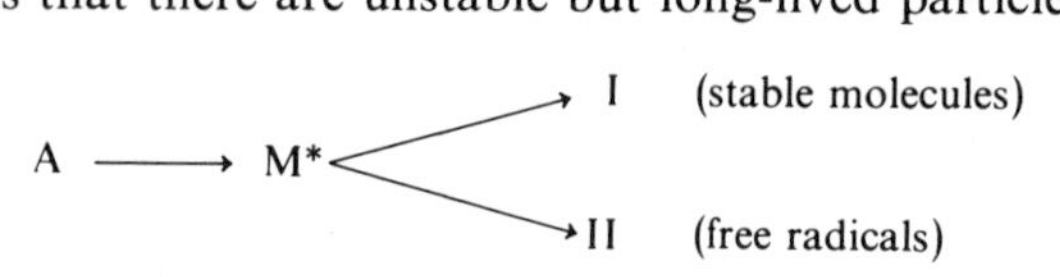

Reaction I is controlled by a high activation energy process and reaction II is controlled by a low activation energy process.

Figure 5 shows the domination of I or II as a function of temperature. As the temperature along reaction route II are raised, more radicals are created and the system passes from steady region 1 to explosive region 2.

* Designates unstable molecule.

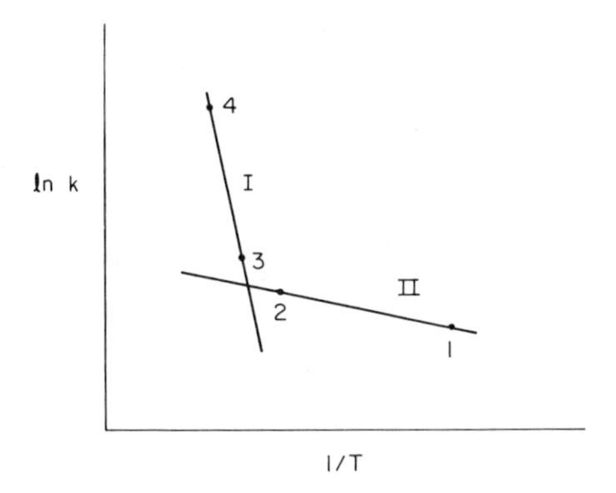

Fig. 5. An Arrhenius plot of the Semenov steps in hydrocarbon oxidation. Points 1–4 correspond to the same points on Fig. 4.

Further raising of the temperature causes a shift to reaction route I which gives stable molecules and thus steady state reaction. Thus since there are no radicals, the system moves into region 3. Raising temperatures along route I still further leads to a reaction so fast that it becomes self-heating and thus explosive again (region 4) (see Chapter 7).

The temperature domination, as shown above, explains the peninsula in the *P–T* diagram and a negative coefficient of reaction rate appears because of the shift from region 2 to 3.

The cool flame phenomenon is a result of the type of experiment performed to determine the explosion limits and the negative temperature coefficient feature of the explosion limits. Most experiments are performed in vessels contained in isothermal liquid baths, as described in the discussion of H_2–O_2 explosion limits. Such systems are considered to be isothermal within the vessel itself. However, the cool gases that must enter the vessel will become hotter at the walls of the vessel than in the center. Reaction starts at the walls and then propagates to the center. However, the initial reaction volume, which is the outermost shell of gases in the vessel, is quickly heated due to the exothermicity of the reaction. This zone immediately moves into the steady reaction area, and the volume does not proceed to explosive reaction and thus to complete and almost instantaneous consumption of reactants. Each successive inner (shell) zone is initiated by the previous zone and proceeds in the same manner. Since there is some chemiluminescence during the initial reaction stages, it appears as if a flame propagates through the mixture. Indeed, the events that occur meet all the requirements of an ordinary flame, except that the reacting mixture loses its explosive characteristic. Thus there is no chance for the mixture to react completely and reach its adiabatic (flame) temperature. Thus the phenomenon is called "cool" flames.

After the complete vessel moves into the slightly higher temperature steady reaction zone, it begins to be cooled by the liquid bath. The mixture temperature drops, and the phenomena can repeat itself a number of times according to the specific experimental conditions and mixtures being studied.

In 1947, Walsh proposed a mechanism which leads to the M* suggested by Semenov. This mechanism which has become accepted, particularly for low temperature oxidation, would be applicable to the higher-order hydrocarbons. Walsh proposed the hydroperoxide for the metastable intermediate. The general form of this hydroperoxide formation is written as

$$\begin{array}{ccccccc} & & H & & H & & \\ & & | & & | & & \\ R & - & C & - & C & - & H \\ & & | & & | & & \\ & & H & & H & & \end{array} \xrightarrow{+O_2} \begin{array}{ccccc} & & H & & H \\ & & | & & | \\ R & - & C & - & CH \\ & & | & & | \\ & & O & & H \\ & & | & & \\ & & O & & \\ & & | & & \\ & & H & & \end{array}$$

where R is an organic radical. The hydroperoxide group forms at the position where the first hydrogen is abstracted. Generally, the position from which the first hydrogen is abstracted is the second carbon atom for straight chain compounds and the tertiary carbon in "iso"-compounds. Theoretical considerations have shown, in addition, that these are the weakest C—H bonds. In this sense, weakest means a few kilocalories less than similar C—H bonds. At low temperatures, a few kilocalories in bond energies can be of great import; however, at higher temperatures, they are not, and abstraction can take place at random; in fact, other bonds could break and lead to still different mechanisms.

The importance in selecting the hydroperoxide is that it favors branched chain reactions. A further reason for postulating its presence and its importance is that it is easy to explain the presence of aldehydes through its breakdown. If aldehydes were chosen as the intermediate, one could explain neither M* nor the induction period.

Just how does the hydroperoxide occur? A mechanism is one in which an oxygen molecule can abstract a hydrogen atom, form HO_2, and leave a radical. Then, another O_2 molecule can attach itself to the radical and form a peroxide. The complex peroxide can strip hydrogen off another hydrocarbon molecule and form the hydroperoxide and another radical (R). The process that is depicted below is not chain branching as one radical is removed for every one formed.

$$\mathrm{R-\overset{H}{\underset{H}{C}}-\overset{H}{\underset{H}{C}}H + O_2 \longrightarrow R-\overset{H}{\underset{\downarrow O_2}{\dot{C}}}-\overset{H}{\underset{H}{C}}H + HO_2}$$

$$\mathrm{R-\overset{H}{\underset{O-O}{C}}-\overset{H}{\underset{H}{C}}H + RH \longrightarrow R-\overset{H}{\underset{O-O-H}{C}}-CH_3 + R'}$$

With the hydroperoxide metastable molecule, Walsh suggested routes I and II to proceed as depicted below.

$$\mathrm{R(\overset{H}{C}OOH)CH_3} \begin{cases} \xrightarrow{\text{I}} & \underbrace{\mathrm{R(CO)CH_3 + H_2O}}_{\text{stable oxidation step}} \\ \xrightarrow{\text{II}} & \underbrace{\mathrm{R + CH_3CHO + OH}}_{\text{branched chain reaction}} \end{cases}$$

The presence of an aldehyde catalyzes the reaction because of subsequent aldehyde oxidation steps which lead to chain branching and a tendency toward explosion; however, the rate of formation of the hydroperoxide is slow and the overall tendency to runaway disappears more quickly.

Based on the Walsh mechanism, the last leg of the hydrocarbon explosion curve (region 3 to region 4) is a self-heating characteristic coming about from reaction route I, as explained before.

More recent work suggests slightly different alternates to the hydroperoxide alone and gives importance to the peroxyl radical ($R\dot{O}_2$). Although the Walsh mechanism appears to explain the explosion limits, the following basic mechanism based on $R\dot{O}_2$ proposed by Semenov (1958) not only does as well but also more readily explains the presence of known intermediaries.

$$\mathrm{RH + O_2 \xrightarrow{1} \dot{R} + H\dot{O}_2} \qquad \text{chain initiation}$$

$$\left.\begin{aligned}
&\mathrm{\dot{R} + O_2 \xrightarrow{2} R\dot{O}_2} \\
&\mathrm{\dot{R} + O_2 \xrightarrow{3} olefin + H\dot{O}_2} \\
&\mathrm{R\dot{O}_2 + RH \xrightarrow{4} ROOH + \dot{R}} \\
&\mathrm{R\dot{O}_2 \xrightarrow{5} R'CHO + R''O\cdot} \\
&\mathrm{H\dot{O}_2 + RH \xrightarrow{6} H_2O_2 + \dot{R}}
\end{aligned}\right\} \text{chain propagation}$$

$$\left.\begin{aligned}
&\mathrm{ROOH \xrightarrow{7} R\dot{O} + \dot{O}H} \\
&\mathrm{R'CHO + O_2 \xrightarrow{8} R'\dot{C}O + H\dot{O}_2}
\end{aligned}\right. \quad \text{degenerate branching}$$

$$\mathrm{R\dot{O}_2 \xrightarrow{9} destruction} \qquad \text{chain termination}$$

The concept of degenerate branching comes about from the delay in decomposition of the respective species in reactions (7) and (8). Thus, as one radical is used up to form the reactants in reactions (7) and (8), the multiple radicals do not appear until these reactants decompose.

Reaction (1) is the same reaction as discussed before, and it is well to note that this reaction is slow and explains the induction period in hydrocarbon combustion.

Reactions (1) and (8) are of the same type. The first is endothermic by 45–50 kcal and the second by 32–33 kcal. Branching would be of importance only when considerably faster than initiation.

Reaction (2) is of near-zero activation energy and fast. It is important to realize that the main method of forming R radicals after initiation, that is when oxidation is proceeding, must be the attack of radicals ($\dot{H}$, $\dot{C}H_3$, $\dot{O}H$, $H\dot{O}_2$, $R\dot{O}$, $R\dot{O}_2$) on the abundant fuel particles; e.g., reaction (6) or $\dot{O}H + RH \rightarrow H_2O + \dot{R}$. The rate constant for the hydroxyl reaction makes this reaction one of the primary means for producing the fuel radicals.

Reaction (3) leads to the olefins known to occur in the oxidation of saturated hydrocarbons. At higher temperature, the simple decomposition of the radical to a lower olefin can occur. If this reaction is rapid, then the oxidation of the saturated hydrocarbon may convert rapidly to the oxidation of a lower olefin.

Reactions (4) and (5) yield the main oxidation intermediates. For liquid phase oxidation (important for hardening of oil spills) occurring at temperatures below 200°C, reaction (4) would undoubtedly be the main reaction, and branching would occur by reaction (7). At gas phase oxidation temperatures ($\simeq$300°C or higher), it is now believed that this reaction is not important. The important suggestion has been made that decomposition follows isomerization of the radical; the activation energy of this reaction is close to 20 kcal/mole. Thus for gas phase oxidation, the main chain propagating reaction would be that of peroxy radical decompostion (reaction (5)) and branching would, consequently, occur by reaction (8). As an example, consider propane oxidation.

$$H_3C{-}\overset{H}{\underset{\cdot}{C}}{-}CH_3 + O_2 \longrightarrow H_3C{-}\overset{H}{\underset{O-O}{C}}{-}CH_3 \longrightarrow \underset{O\,-\,O}{H_3C{-}\dot{C}H \quad CH_3}$$

$$\longrightarrow CH_3CHO + CH_3\dot{O}$$

It should be pointed out that the angle of the peroxide bonding (—COO) has been established to be 90° and gives credance to the above and similar mechanisms discussed.

The competing reaction scheme between explosive and steady reaction can now be written in the form different from that given previously and one that is more acceptable for the higher temperature gas phase systems of most concern here. For example

$$\underset{\text{(any radical)}}{X} \xrightarrow[+RH]{(a)} \dot{R} \xrightarrow[+O_2]{(b)} R\dot{O}_2 \begin{cases} \xrightarrow{(c)\ \text{isomerization}} \dot{R}\,OOH \xrightarrow[O_2]{(d)} \text{branching chain} \\ \xrightarrow{(c)\ \text{decomposition}} \text{free radical} \longrightarrow \text{straight chain (steady reaction)} \end{cases}$$

where X is one of the free radicals as mentioned in discussing reaction (2). Once the chain is initiated by reaction (1) and the overall reaction system begins to take place, $\dot{R}$ can be formed by reaction (1), but is most likely formed by any of the lower activation energy reactions represented by step (a) above. Step (b) is explicitly reaction (2). Step (d) is written as

$$\cdot ROOH \xrightarrow[+O_2]{} \cdot OOR'{-}R''OOH \xrightarrow[RH]{} HOOR'{-}R''OOH + R$$

$$\longrightarrow \underset{\text{ketone}}{R'''CO} + \underset{\text{aldehyde}}{R^{IV}CHO} + 2\dot{O}H$$

Consider the example of the oxidation of 2-methylpentane. The hydroperoxy state is structured as

$$(CH_3)_2C(OO)CH_2CH_2CH_6 \longrightarrow (CH_3)_2C(O{-}OH)\dot{C}HCH_2CH_3 \xrightarrow{+O_2} (CH_3)_2C(O{-}OH)CH(O{-}O\cdot)CH_2CH_3$$

$$\xrightarrow{RH} (CH_3)_2C(O{-}OH)CH(O{-}OH)CH_2CH_3 \longrightarrow (CH_3)_2CO + CH_3CH_2CHO + 2OH$$

Reaction (3) accounts for the cracking products under oxidation. It has been well established that thermal cracking of pure hydrocarbons takes place much more readily in the presence of small amounts of oxygen. Reaction (3) is a logical step to explain this phenomenon. The resulting allylic hydrogen is weaker and more easily attacked by other radicals. Further, allylic radicals can decompose readily to a di-olefin and a smaller radical or H atom. The oxidation of olefins proceeds quite readily and will be discussed later.

The observance of other products not derivable from intermediate aldehydes, particularly ring ethers, alcohols, and acids, indicates that other fates for peroxy radicals are possible. For example,

```
H  H  H  H                      H  H  H  H
HC—C—C—C—R     ——→      HC—C—C—C—R + OH
H  H  H  |                      H  |  H  |
      O—O                          └─O─┘
```

where the O internally abstracts the hydrogen atom from the second carbon atom from the other end.

The alcohol follows from the products of reaction (5) or (7),

$$R\dot{O} + XH \longrightarrow ROH + X$$

Acids, ketones, etc., can form from various possible cleavages of bonds, as represented in the discussion of 2-methylpentane.

Thus, it is possible to see that a wide range of intermediate compounds could form in small concentrations during the low temperature oxidation of hydrocarbons. Thus, in cold flammable mixtures brought to combustion, such as in automotive engines, diesel, flames, etc., it is not surprising that most of these compounds have been identified. Again, it is worth repeating that the concentrations of these species are so low they can be ignored in thermodynamic calculations.

The hydrocarbon oxidation that has been discussed to this point is normally classified as "low temperature" combustion. It should be noted further that only the higher order paraffin hydrocarbons have been considered. At very high temperatures, explosive conditions exist at essentially all pressures but, more importantly, one must realize, as will be explained in succeeding sections, that a high temperature cracking (or pyrolysis) of the hydrocarbon to radicals and subsequent reactions, including the steps in the H_2–O_2 reaction, can occur and explain the chain branching necessary for explosion. Thus, there are fundamentally three regions of hydrocarbon oxidation: steady reaction, low temperature (explosive) combustion, and high temperature (explosive) combustion. A premixed fuel–oxygen (or air) mixture at room temperature (or slightly higher), if reacting at all, is reacting ever so slightly. For a flame to propagate, the reaction rates must be fast; i.e., explosive. Thus an ignition source is used to bring the low temperature mixture into the higher temperature explosive regime. Energy release brings the system into the highest temperature combustion kinetics regime. As the flame travels, the temperature of the mixture rises from room temperature to the adiabatic flame temperature. Thus the gases pass through all three hydrocarbon oxidation regimes. As will be discussed more fully in the section on flames, it is the high temperature regimes which determine the flame speed. To reiterate, however, quenching or cooling effects in practical combustion systems bring the reacting mixture into the regime discussed in the previous pages. Except for NO_x and CO, most pollutants form in the lower temperature oxidation regime.

3. Detailed Oxidation Mechanisms of Hydrocarbons

a. Methane

Methane exhibits certain oxidation characteristics which are different from all other hydrocarbons. Tables of bond energy list the first broken C—H bond in methane to be kilocalories more than the others, and certainly more than the C—H bond on the second carbon in a longer chain hydrocarbon. Thus it is not surprising to find various kinds of experimental evidence to lead one to believe that it is more difficult to ignite methane/air (oxygen) mixtures than it is other hydrocarbons. At low temperatures, even oxygen radical attack is slow. Indeed, in discussing exhaust emissions with respect to pollutants, the terms total hydrocarbons and reactive hydrocarbons are used. The difference between the two terms is simply methane, which in this context is considered to react so slowly with oxygen radical in the atmosphere that it is called unreactive.

Since methane contains only one carbon atom, it does not form acetaldehyde, which readily gives chain branching even under low temperature combustion conditions. It does form formaldehyde, which provides the chain branching step necessary for explosion.

The simplest scheme which will explain the lower temperature results of methane oxidation is

Reaction		
$CH_4 + O_2 \xrightarrow{1} CH_3 + HO_2$	chain initiation
$\dot{C}H_3 + O_2 \xrightarrow{2} CH_2O + \dot{O}H$	chain propagation
$OH + CH_4 \xrightarrow{3} H_2O + \dot{C}H_3$	chain propagation
$OH + CH_2O \xrightarrow{4} H_2O + HCO$	chain propagation
$CH_2O + O_2 \xrightarrow{5} HO_2 + HCO$	chain branching
$H\dot{C}O + O_2 \xrightarrow{6} CO + HO_2$	chain propagating
$HO_2 + CH_4 \xrightarrow{7} H_2O_2 + CH_3$	chain propagating
$H\dot{O}_2 + CH_2O \xrightarrow{8} H_2O_2 + H\dot{C}O$	chain propagating
$OH \xrightarrow{9} \text{wall}$	chain termination
$CH_2O \xrightarrow{10} \text{wall}$	chain termination

As before, reaction (1) is slow. Reactions (2) and (3) are fast since they involve a radical and one of the initial reactants. The same is true for reactions (5)–(7). Reaction (5) represents the necessary chain branching step. Reactions (4) and (8) introduce the formyl radical known to exist in the low-temperature combustion scheme.

The combustion of methane at higher temperature involves the further oxidation of CO to CO_2. In hydrocarbon systems, one can tell whether a particular reference is discussing low temperature oxidation or not, by observing the fate of CO. If CO conversion to CO_2 is not written, then the system of rate expressions is for low temperature. The conversion of CO to CO_2 competes with the direct oxidation of CH_4 and retards it. It has been established that the reaction

$$OH + CO \longrightarrow H + CO_2$$

competes with reaction (3) and is only an order of magnitude or less slower. Thus, when the concentration of CO builds up, the effect of the above step is great.

At higher temperatures, certain high activation energy steps become feasible, particularly the pyrolysis reactions. The presence of larger concentrations of O and H radicals must be felt as well. A complete system of equations and approximate rate constants have been given by Seery and Bowman (1970) and are shown in Table 1.

It has been determined by Seery and Bowman that at temperatures of the order of 2000°K, reaction (1) is the only one of the first three radical

TABLE 1

	Reaction	k_f, rate constant
1.	$CH_4 + M \rightarrow CH_3 + H + M$	$1.5 \times 10^{19} \exp(-100{,}600/RT)$
2.	$CH_4 + O_2 \rightarrow CH_3 + HO_2$	$1.0 \times 10^{14} \exp(-45{,}400/RT)$
3.	$O_2 + M \rightarrow 2O + M$	$3.6 \times 10^{18} T^{-1.0} \exp(-118{,}800/RT)$
4.	$CH_4 + O \rightarrow CH_3 + OH$	$1.7 \times 10^{13} \exp(-8760/RT)$
5.	$CH_4 + H \rightarrow CH_3 + H_2$	$6.3 \times 10^{13} \exp(-12{,}700/RT)$
6.	$CH_4 + OH \rightarrow CH_3 + H_2O$	$2.8 \times 10^{13} \exp(-5000/RT)$
7.	$CH_3 + O \rightarrow H_2CO + H$	10^{13}–10^{15}
8.	$CH_3 + O_2 \rightarrow H_2CO + OH$	10^{11}–10^{14}
9.	$H_2CO + OH \rightarrow HCO + H_2O$	10^{13}–10^{15}
10.	$HCO + OH \rightarrow CO + H_2O$	10^{12}–10^{15}
11.	$CO + OH \rightarrow CO_2 + H$	$3.1 \times 10^{11} \exp(-600/RT)$
12.	$H + O_2 \rightarrow OH + O$	$2.2 \times 10^{14} \exp(-16{,}600/RT)$
13.	$O + H_2 \rightarrow OH + H$	$4.0 \times 10^{14} \exp(-9460/RT)$
14.	$O + H_2O \rightarrow 2OH$	$8.4 \times 10^{14} \exp(-18{,}240/RT)$
15.	$H + H_2O \rightarrow H_2 + OH$	$1.0 \times 10^{14} \exp(-20{,}400/RT)$
16.	$H + OH + M \rightarrow H_2O + M$	$2.0 \times 10^{1} T^{-1.0}$
17.	$CH_3 + O_2 \rightarrow HCO + H_2O$	10^{11}–10^{12}
18.	$HCO + M \rightarrow H + CO + M$	$2.0 \times 10^{13} T^{1/2} \exp(-28{,}800/RT)$

initiation reactions fast enough to explain ignition results by shock waves. The attack of CH_4 by various radicals formed later in the system is shown in reactions (4)–(6). From considerations of concentrations, OH and H would dominate and the effectiveness of the low activation OH reaction is appreciated. In certain kinetic situations, one must be careful in neglecting reaction (4). At high temperatures in methane and all other hydrocarbon systems, the primary initiation step is almost certainly the pyrolysis of the hydrocarbon molecule to give alkyl radicals.

Various flame studies have shown that the methyl radical is attacked by O or O_2 forming formaldehyde, as shown in reactions (7) and (8). At high temperatures ($\sim 1000°C$), Benson (1972) has ruled out reaction (8) on thermodynamic grounds. Reaction (7) must proceed via the route $M + CH_3 + O \rightarrow CH_3O + M \rightarrow CH_2O + H$. The formaldehyde steps follow in reactions (9) and (10).

During the induction period when $(O)/(O_2) \ll 10^{-3}$, Seery and Bowman state that reaction (7) may be neglected in comparison with reaction (8); however, this statement was made before knowledge of Benson's calculations.

For $k_9/k_8 > 10^2$, reactions (8) and (9) can be combined

$$CH_3 + O_2 \longrightarrow HCO + H_2O$$

to give reaction (17).

The other point worth noting is that formyl decomposition (reaction (18)) can become important as well. It is not surprising from the high and lower temperature mechanisms that the addition of formaldehyde reduces the induction period of methane oxidation at both high and low temperatures. Notice, as well, that the HO_2 radical is not included in the high temperature scheme. Its lifetime at these temperatures is just too short.

The CO oxidation step via OH, and the H_2–O_2 reaction steps must be included because reaction (11) produces H radicals and the concentration of O radicals is also greater at the high temperature. None of these steps is then appropriate for the low temperature oxidation.

b. Aldehydes

There has been frequent mention to the point that aldehydes reduce the induction period and contribute to the explosion condition by providing the chain branching step. In the various mechanisms given already, the elements of acetaldehyde and formaldehyde oxidation have essentially been discussed.

One of the degenerate branching steps in the discussion of paraffin hydrocarbon oxidation was

$$R'CHO + O_2 \longrightarrow R'CO + HO_2$$

If R′ were a methyl radical, then the fuel is acetaldehyde. Indeed

$$CH_3CHO + O_2 \longrightarrow CH_3\dot{C}O + HO_2$$

is the postulated initiation step in acetaldehyde oxidation. Acetyl radicals may be formed as well from such steps with radical X as

$$CH_3CHO + X \longrightarrow CH_3\dot{C}O + XH$$

That the radical $CH_3\dot{C}O$ forms is supported by deuterated aldehyde studies that show that the weakest C—H bond in aldehydes is that in the aldehyde group.

The subsequent step is the decomposition of the $CH_3\dot{C}O$ radical

$$CH_3\dot{C}O \longrightarrow \dot{C}H_3 + CO$$

As discussed in the previous section on methane oxidation, depending on the temperature, the methyl radical is oxidized to formaldehyde (H_2CO) through one of the following routes

$$CH_3 + O \longrightarrow H_2CO + H$$

$$CH_3 + O + M \longrightarrow H_3CO + M \longrightarrow H_2CO + H + M$$

$$CH_3 + O_2 \longrightarrow H_2CO + OH$$

Following the previous section, it is evident that the formaldehyde is oxidized through the steps

$$CH_2O + OH \longrightarrow H_2O + HCO$$

$$CH_2O + O_2 \longrightarrow HO_2 + HCO$$

$$CH_2O + HO_2 \longrightarrow H_2O_2 + HCO$$

$$HCO + O_2 \longrightarrow CO + HO_2$$

$$HCO + OH \longrightarrow H_2O + CO$$

$$HCO + M \longrightarrow H + CO + M$$

At high temperatures the HO_2 is not stable, and steps in which it is involved are not found as written above. Similarly, at low temperatures hydroxyl concentrations are generally not high, and the next to last reaction is not likely since it involves two radicals as reactants. The last reaction is a high temperature step.

c. *Higher Paraffin Hydrocarbons*

Combustion of paraffins above methane is complicated by the greater instability of the higher alkyl radicals and by the great variety of secondary products which can form. The mechanism characteristically follows the

Semenov type discussed earlier. Detailed oxidation mechanisms of some hydrocarbons can be found in the literature (Minkoff and Tipper, 1962).

At higher temperatures, most have accepted the primary reaction to be between the hydroxyl radical and the fuel:

$$\mathrm{RH + OH \longrightarrow \dot{R} + H_2O}$$

The initiation step, of course, is the decomposition of the hydrocarbon molecule to give an alkyl radical. Recent work has suggested that other reactions in addition to the OH attack were important; namely, in fuel lean and rich combustion:

$$\mathrm{RH + O \longrightarrow \dot{R} + OH}$$

and in fuel rich combustion:

$$\mathrm{RH + H \longrightarrow R + H_2}$$

These reactions serve as well to emphasize the importance of the chain steps (reactions (12)–(14) in the methane mechanism discussed in Section 3a) in all hydrocarbon oxidation processes.

It is interesting to review a general pattern for the oxidation of hydrocarbons in flames as suggested by Fristrom and Westenberg (1965). They suggest two essentially thermal zones: the primary zone in which the initial hydrocarbons are attacked and reduced to CO, H_2, H_2O and the various radicals (H, O, OH) and the secondary zone in which the CO and H_2 are oxidized. The primary zone, of course, is where the intermediates occur. In oxygen-rich saturated hydrocarbon flames, they suggest that initially hydrocarbons lower than the initial fuel form according to

$$\mathrm{OH + C_nH_{2n+2} \longrightarrow H_2O + C_nH_{2n+1} \longrightarrow C_{n-1}H_{2n-2} + CH_3}$$

Because hydrocarbon radicals higher than ethyl are thought to be unstable, the initial radical C_nH_{2n+1} usually splits off CH_3 and forms the next lower olefinic compound, as shown. With hydrocarbons higher than C_3H_8, it is thought there may be fission into an olefinic compound and a lower radical. The radical alternatively splits off CH_3. The formaldehyde which forms in the oxidation of the fuel and radicals is rapidly attacked in flames by O, H, and OH, so that formaldehyde is usually only found as a trace in flames.

In fuel-rich saturated hydrocarbon flames, Fristrom and Westenberg state the situation is more complex, although the initial reaction is simply the H atom abstraction analogous to the preceding OH reaction; e.g.,

$$\mathrm{H + C_nH_{2n+2} \longrightarrow H_2 + C_2H_{2n+1}}$$

Under these conditions the concentration of H and other radicals are large enough that their recombination becomes important and hydrocarbons higher than the original fuel are formed as intermediates.

The general features suggested by Fristrom and Westenberg have been confirmed by Glassman *et al.* (1975) in high temperature flow reactor studies. However, this new work permits more detailed understanding of the high temperature oxidation mechanism. As stated earlier, this work shows that under oxygen-rich conditions initial attack by O atoms must be considered as well as the primary OH attack. More importantly, however, it has been established that the paraffin reactants produce intermediate products which are primarily olefinic, and the fuel is consumed to a major extent before significant energy release occurs. The higher the initial temperature, the greater the energy release as the fuel is being converted. This observation leads one to conclude that the olefin oxidation rate simply increases more appreciably with temperature; i.e., the olefins are being oxidized while they are being formed from the fuel.

Analyses of the intermediates formed from some paraffin hydrocarbons are given in Table 2. It would appear that the results given in Table 2

TABLE 2

Relative importance of intermediates in hydrocarbon combustion

Fuel	Relative hydrocarbon intermediate concentrations
ethane	ethene $\gg$ methane
propane	ethene $>$ propene $\gg$ methane $>$ ethane
butane	ethene $>$ propene $\gg$ methane $>$ ethane
hexane	ethene $>$ propene $>$ butene $>$ methane $\gg$ pentene $>$ ethane
2-methyl pentane	propene $>$ ethene $>$ butene $>$ methane $\gg$ pentene $>$ ethene

would contradict Fristrom and Westenberg's suggestion that the initial hydrocarbon radical C_nH_{2n+1} usually splits off the methyl radical. If this type of splitting were to occur, one could expect to find larger concentrations of methane. The large concentrations of ethene found in all cases would suggest that primarily the initial C_nH_{2n+1} radical cleaves one bond from the carbon atom from which the hydrogen was abstracted. The bond next to this carbon atom is less likely to break since this type of cleavage would require both an electron and hydrogen transfer to form the olefin. The abstraction of hydrogen from a second carbon atom requires about 1.5 kcal less from the other carbon atoms (a tertiary carbon atom requires about 2.5 less). In a straight chain hydrocarbon there are, of course, more hydrogens on the first carbon atoms. Estimating relative probability of removal based on number and ease of removal and considering the cleavage rule mentioned indicates the proper trends designated by Table 2 and relatively large concentrations of ethene and propene. These results suggest that oxidation studies of ethene and propene should be particularly important.

The evidence is thus that there are three distinct but coupled zones in hydrocarbon combustion.

1. Following ignition, primary fuel disappears with little or no energy release and produces unsaturated hydrocarbons and hydrogen. A little of the hydrogen is concurrently being oxidized to water.
2. Subsequently, the unsaturated compounds are further oxidized to carbon monoxide and hydrogen. Simultaneously the hydrogen present and formed is oxidized to water.
3. Last, the large amounts of carbon monoxide formed are oxidized to carbon dioxide and most of the heat release from the primary fuel is obtained.

The cycloparaffins are thought to follow the same type of low temperature mechanism as the straight chain hydrocarbons and to produce cyclic organic compounds as intermediates. In light of the new high temperature results on the straight and branched chained hydrocarbon, it is difficult to speculate on what the mechanism of the high temperature oxidation of cycloparaffins would be.

d. Olefins

Ethylene is rapidly attacked by O atoms, methyl, OH or other radicals. The attack is usually an addition to the double bond rather than hydrogen abstraction, the O attack resulting in the formation of ethylene oxide. The major steps, as before, for low temperatures are

$$HO_2 + C_2H_4 \longrightarrow \underset{\diagdown\,O\,\diagup}{\overset{H\ \ H}{HC{-}CH}} + OH \longrightarrow CH_3{-}CHO + OH$$

Recall

$$OH + C_2H_4 \longrightarrow \underset{H-O}{\overset{H\ \ H}{HC{-}CH}} \longrightarrow CH_3 + CH_2O$$

$$CH_3 + O_2 \longrightarrow CH_2O + OH$$

CH_2O can undergo the characteristic reactions discussed previously. In flames, it appears again that primary attack is by OH radicals with the resulting fragment reacting with O_2 as follows:

$$OH + C_2H_4 \longrightarrow H_2O + C_2H_3$$

$$C_2H_3 + O_2 \longrightarrow H_2CO + HCO$$

H H
| |
HC—C
   |
O—O

e. Aromatics

The appropriate mechanism for benzene oxidation appears to be

$$C_6H_6 + O_2 \longrightarrow \dot{C}_6H_5 + H\dot{O}_2$$

$$\dot{C}_6H_5 + O_2 \longrightarrow C_6H_5O\dot{O}$$

$$C_6H_5O\dot{O} + C_6H_6 \longrightarrow C_6H_5\dot{O} + C_6H_5 + OH$$

$$C_6H_5\dot{O} + C_6H_6 \longrightarrow C_6H_5OH + \dot{C}_6H_5$$

$$C_6H_5OH \xrightarrow{\text{as above}} C_6H_4(OH)_2$$

A reaction of $H\dot{O}_2$ with benzene to give H_2O_2 is ruled out since hydrogen peroxide has not been detected in benzene oxidation. Basically the oxidation of benzene does not differ from those of other hydrocarbons. The induction periods are longer because of the greater difficulty in abstracting the first hydrogen. The phenol which forms in high yield is due to its great stability, and it is certainly possible to get many cyclic intermediates which could lead even to acetylene formation. For example,

OH, OH (hydroquinone) $\xrightarrow{+O_2}$ OH, C, HC, O, CH, HC, O, CH, C, OH $\longrightarrow$ OH, C=O, HC, HC, O, C, OH $+ \; C_2H_2$ acetylene

The oxidation of acetylene will give formaldehyde and formic acid and may lead to chain branching.

REFERENCES

Benson, S. W. (1972). *Nat. Bur. Std. Spec. Publ.* 357, p. 121.
Brokaw, R. S. (1967). *Int. Symp. Combust., 11th* p. 1063. Combustion Inst. Pittsburgh, Pennsylvania.

Fristrom, R. M., and Westenberg, A. A. (1965). "Flame Structure," Chapter XIV. McGraw-Hill, New York.

Glassman, I., Dryer, F. L., and Cohen, R. (1975). *Int. Symp. Chem. Reaction Dynam., 2nd* Univ. of Padua, Padua, Italy.

Gordon, A., and Knipe, R. (1955). *J. Phys. Chem.* **59**, 1160.

Lewis, B., and von Elbe, G. (1961). "Combustion, Flames and Explosions in Gases," 2nd ed., Part 1. Academic Press, New York.

Minkoff, C. J., and Tipper, C. F. H. (1962). "Chemistry of Combustion Reactions." Butterworths, London.

Seery, D., and Bowman, C. T. (1970). *Combust. Flame* **14,** 37.

Semenov, N. N. (1958). "Some Problems in Chemical Kinetics and Reactivity," Chapter VII. Princeton Univ. Press, Princeton, New Jersey.

Chapter Four

Flame Phenomena in Premixed Combustible Gases

Figure 1 depicts the general characteristics of the flame anchored on top of a simple Bunsen burner. Recall that the fuel gas entering a laboratory Bunsen burner induces air from the surroundings. As the fuel and air flow up the tube they mix, and before the top of the burner the mixture is completely homogeneous. The dark zone designated in Fig. 1 is simply the premixed gases before they enter the luminous zone where reaction and heat release are taking place. The luminous zone is no more than 1 mm thick and, in fact, is usually less. More specifically, the luminous zone is that portion of the reacting zone in which the temperature is the highest and indeed is the zone in which most of the reaction is taking place. The color of the luminous zone changes with the fuel–air ratio. When the mixture is fuel-lean, a deep violet radiation due to excited CH radicals appears. When the mixture is fuel-rich, the radiation is green and is due to the C_2 molecule. The high temperature burned gases usually show a reddish glow which arises from radiation from CO_2 and water vapor. When the mixture is adjusted to be very fuel-rich, carbon particles form and an intense yellow radiation appears. This radiation is continuous, and is due to the presence of the solid carbon particles. The appearance is yellow because Planck's black body curve peaks in the yellow for the temperatures that normally exist in these flames.

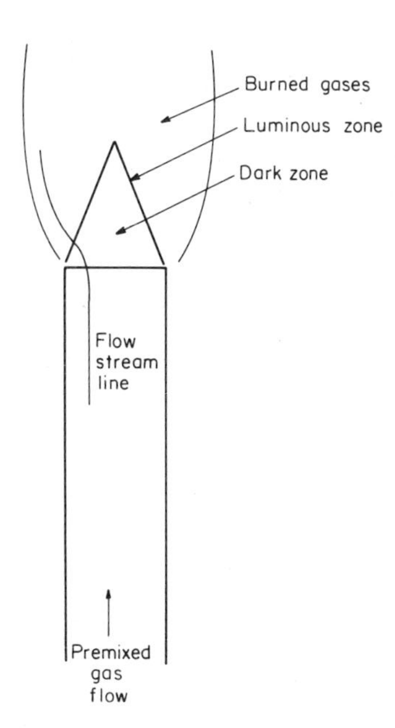

Fig. 1. A characteristic Bunsen burner flame.

With the background of the hydrocarbon oxidation kinetics given in the previous chapter, it is possible to characterize, as Bradley (1969) has done, the narrow reaction region into three zones: a preheat zone, a reaction zone, and a recombination zone. The nature of the reactions in the preheat zone depends very much on the fuel involved. For a very stable molecule like methane, little or no pyrolysis can occur within the short residence time in the flame. With the majority of the other saturated hydrocarbons, considerable degradation occurs, and the fuel fragments which leave this zone consist of mainly lower olefins, hydrogen, and lower hydrocarbons. Thus the composition in the reaction zone is very similar, irrespective of the nature of the fuel. Since the flame temperature of the saturated hydrocarbons would also be very nearly the same for reasons discussed in Chapter 1, it is not surprising then that their burning velocities, which will be shown to be very dependent on reaction of hydrocarbons, would all be of the same rates order (~ 40 cm/sec).

In the preheat zone, the oxygen acts as a homogeneous catalyst for the pyrolysis of the hydrocarbon and little of the oxygen is consumed. The oxidation steps occurring in the reaction zone are those which were discussed in detail in the previous chapter. The recombination zone falls into the burned gas or post-flame zone. It is a more extended region in which the much slower recombination reactions occur.

The dotted line in Fig. 1 depicts a flow streamline. If one measures the temperature along the streamline, it is found that the dark zone temperature is that of the unburned gases T_0. As the reaction zone is approached, the temperature rises exponentially and approaches the adiabatic flame temperature T_f just past the luminous zone. Fig. 2 depicts such a temperature trace.

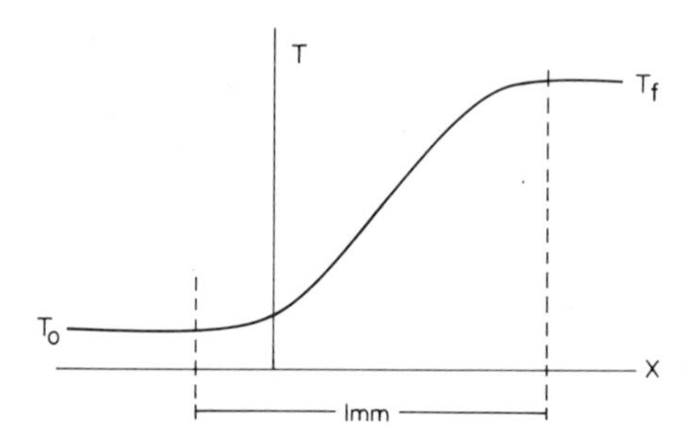

Fig. 2. A temperature profile through a laminar Bunsen flame.

It is interesting to consider what would occur if a gas mixture of the same composition as that which occurs in an ideal Bunsen burner were to be stored in a horizontal tube and ignited. First consider the condition in which this mixture is held in the tube by closures at both ends. Then both ends are opened simultaneously and an ignition source applied at one end. A flame appears and propagates through the tube at a speed of the order of 40 cm/sec. The flame can be considered very much like a wave. Indeed the thickness of this wave or flame is about the same as that on the Bunsen burner and, in fact, the temperature profiles through these two flames are the same. Although an actual flame in this tube experiment would not be perfectly flat due to certain buoyancy effects, ideally it may be so treated. If the flame travels from right to left, i.e., the ignition source were placed at the right end of the tube, and a coordinate system traveling with the flame is chosen, then the unburned gases enter from the left to the right, and the burned gases are to the right of the flame as shown in Fig. 3. The direction in which the burned gases flow will be discussed later.

Now, essentially the same experiment is repeated except that the ignition end is kept closed. Again, a flame or a wave propagates, except that the velocity of this wave is not tens of centimeters per second, but thousands

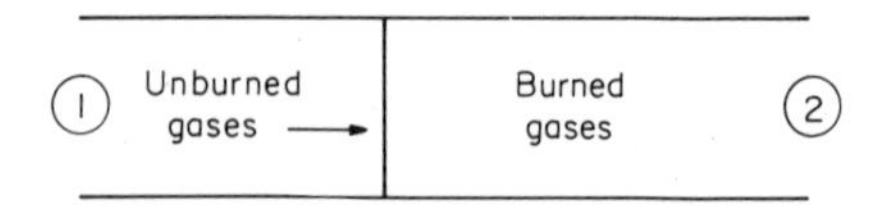

Fig. 3. A combustion wave fixed with respect to a tube.

of meters per second. The latter wave travels at speeds which are supersonic with respect to the unburned gases, and the initial wave travels at speeds which are subsonic with respect to the unburned gases. The fact that both a subsonic and supersonic solution appear to exist leads one to believe that one could approach the problem as if it were purely fluid mechanical in nature. The tendency is to set down the integrated conservation equations and to determine whether solutions exist. This conceptual approach follows.

If the subscript 1 specifies the unburned gas conditions and subscript 2 the burned gas conditions, the equations generally written are

continuity $$\rho_1 u_1 = \rho_2 u_2 \tag{1}$$

momentum $$p_1 + \rho_1 u_1{}^2 = p_2 + \rho_2 u_2{}^2 \tag{2}$$

energy $$c_p T_1 + \tfrac{1}{2} u_1{}^2 + q = c_p T_2 + \tfrac{1}{2} u_2{}^2 \tag{3}$$

state $$p_1 = \rho_1 R T_1 \tag{4}$$

state $$p_2 = \rho_2 R T_2 \tag{5}$$

Equation (4) connects the known variables, unburned gas pressure, temperature and density, and thus is not an independent equation. In the coordinate system chosen, u_1 is the velocity fed into the wave and u_2 is the velocity coming out of the wave. In the laboratory coordinate system, the velocity ahead of the wave is zero, the wave velocity is u_1, and $(u_1 - u_2)$ is the velocity of the burned gases with respect to the tube. The unknowns in the system are u_1, u_2, ρ_2, T_2 and p_2. q is the chemical energy release and T_2 is the stagnation adiabatic combustion temperature.

Notice that there are five unknowns and only four equations. Nevertheless, one can proceed by analyzing the equations at hand. Simple algebraic manipulation (detailed in Chapter 5) results in two new equations

$$\frac{\gamma}{\gamma - 1}\left(\frac{p_2}{\rho_2} - \frac{p_1}{\rho_1}\right) - \frac{1}{2}(p_2 - p_1)\left(\frac{1}{\rho_1} + \frac{1}{\rho_2}\right) = q \tag{6}$$

and

$$\gamma M_1{}^2 = \left(\frac{p_2}{p_1} - 1\right) \Big/ \left[1 - \frac{(1/\rho_2)}{(1/\rho_1)}\right] \tag{7}$$

γ is the ratio of specific heats and M_1 is the wave velocity divided by $(\gamma R T_1)^{1/2}$, or the Mach number of the wave. For simplicity the specific heats are assumed constant, i.e., $c_{p_1} = c_{p_2}$.

Equation (6) is referred to as the Hugoniot relationship and states for given initial conditions $(p_1, 1/\rho_1, q)$ a whole family of solutions $(p_2, 1/\rho_2)$ is possible. The family of solutions lie on a curve in a plot of P_2 versus $1/\rho_2$ as shown in Fig. 4. Plotted on the graph represented by Fig. 4 is

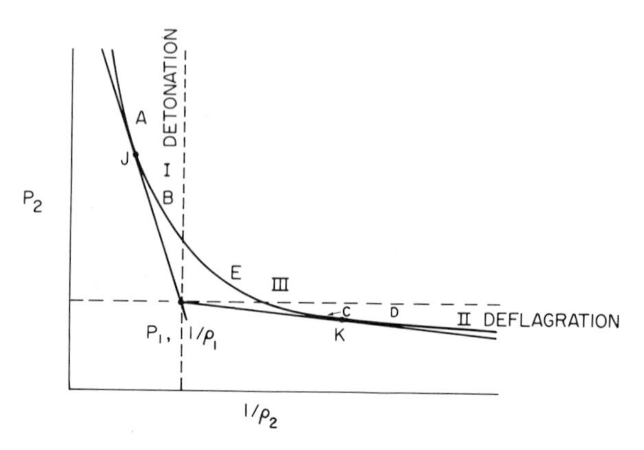

Fig. 4. A Hugoniot plot divided into five regions A–E.

the initial point (p_1, $1/\rho_1$) and the two tangents through this point to the curve representing the family of solutions. One obtains a different curve for different fractional values of q. Indeed, a curve is obtained for $q = 0$; i.e., no energy release. This curve goes through the point representing the initial condition and is referred to as the shock Hugoniot since it gives the solution for simple shock waves.

A horizontal line and a vertical line are also drawn through the initial condition point. These lines, of course, represent the conditions of constant pressure and constant specific volume $1/\rho$, respectively. They further break the curve into three sections. Sections I and II are further divided into sections by the tangency points (J and K) and the other letters defining particular points.

Examination of Eq. (7) reveals the character of M_1 for regions I and II. In region I, P_2 is much greater than P_1, and thus the difference is a number much larger than 1. Further, in this region $1/\rho_2$ is a little less than $1/\rho_1$, and thus the ratio is a number close to but a little less than 1. Therefore the denominator is very small, much less than 1. Consequently the right-hand side of Eq. (7) is very much larger than 1 and certainly greater than 1.4. Conservatively one assumes $\gamma = 1.4$, then $M_1{}^2$ and M_1 are greater than 1. Thus region I gives supersonic waves and is called the detonation region. A detonation can then be defined as a supersonic wave supported by energy release (combustion).

Similarly in region II since P_2 is a little less than P_1, the numerator of Eq. (7) is a small negative number less than 1. $1/\rho_2$ is much greater than $1/\rho_1$, and thus the denominator is a negative number greater than 1. The right-hand side of Eq. (7) for region II is less than 1, consequently M_1 is less than 1. Thus region II gives subsonic waves and is called the deflagration region. Deflagration waves are defined as subsonic waves supported by combustion.

In region III, $P_2 > P_1$ and $1/\rho_2 > 1/\rho_1$; the numerator of Eq. (7) is positive and the denominator is negative. Thus M_1 is imaginary in region III and, therefore, does not represent a physically real solution.

It will be shown in Chapter 5 that the velocity of sound in the burned gases for points on the Hugoniot higher than J is greater than the velocity of the detonation wave relative to the burned gases. Consequently in any real physical situation in a tube, wall effects cause a rarifaction. These rarifaction waves will catch up to the detonation front, reduce the pressure, and cause the final values of P_2 and $1/\rho_2$ to drop to point J, the so-called Chapman–Jouguet point. Points in region B are eliminated by consideration of the wave structure. Thus the only steady state solution in region II is given by point J. This unique solution has been found strictly by fluid dynamic and thermodynamic considerations.

Further, the velocity of the burned gases at J and K can be shown to equal the sound speed there and thus $M_2 = 1$ is a condition at both J and K. An expression similar to Eq. (7) for M_2 reveals that M_2 is greater than 1 as values past K are assumed. Such a condition cannot be real for it would mean that the velocity of the burned gases would increase by heat addition. It is well known that it is not possible to increase the velocity of sonic flow in a constant area duct by heat addition. Thus region D is ruled out. Unfortunately there are no means by which to reduce the range of solutions which is given by region C. In order to find a unique deflagration velocity for a given set of initial conditions, another equation must be obtained. This equation comes about from the examination of the structure of the deflagration wave and deals with the rate of chemical reaction or more especially, the rate of energy release.

The Hugoniot curve shows that in the deflagration region the pressure change is very small. Indeed, approaches seeking the unique deflagration velocity assume the pressure to be constant and drop the momentum equation.

The gases that flow in the Bunsen tube are laminar. Since in the horizontal tube experiment, the wave created is so very similar to the Bunsen flame, it too is laminar. The deflagration velocity under these conditions is called the laminar flame velocity, and it is the subject of laminar flame propagation that is treated in the remainder of this section.

For those who have not studied fluid mechanics, the definition of a deflagration as a subsonic wave supported by combustion may sound oversophisticated, nevertheless it is the only precise definition. Others describe flames in a more relative context. A flame can be considered a rapid, self-sustaining chemical reaction occurring in a discrete reaction zone. Reactants may be introduced into this reaction zone or the reaction zone may move into the reactants depending upon whether the unburned gas velocity is greater than or less than the flame (deflagration) velocity.

A. THE LAMINAR FLAME SPEED

The flame velocity, which is also called the burning velocity, normal combustion velocity, or laminar flame speed must be more precisely defined as the velocity at which unburned gases move through the combustion wave in the direction normal to the wave surface.

The theoretical approaches to the calculation of the laminar flame speed can be divided into three categories:

(a) thermal theories,
(b) diffusion theories,
(c) comprehensive theories.

The historical development followed approximately the same order.

The thermal theories date back to Mallard and Le Chatelier (1885) who proposed that it is propagation of heat back through layers of gas that is the controlling mechanism in flame propagation. And, as one would expect a form of the energy equation is the basis for the development of the thermal theory. Mallard and Le Chatelier (1885) postulated that the flame zone consists of two zones broken up by the point where the next layer ignites, as shown in Fig. 5.

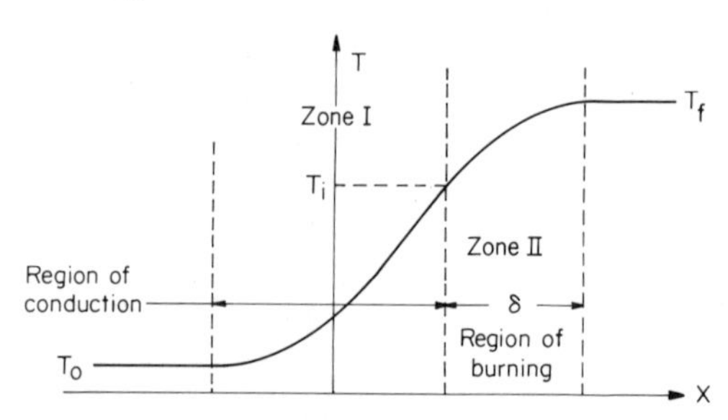

Fig. 5. The Mallard–Le Chatelier description of a laminar temperature wave.

Unfortunately, this thermal theory requires the concept of an ignition temperature and no adequate means exist for the determination of ignition temperatures.

Later, there were improvements in the thermal theories, probably the most significant of which is that due to Zeldovich and Frank-Kamenetskii, whose derivation was presented in detail by Semenov (1951) and is commonly called the Semenov equation. These authors included the diffusion of molecules as well as heat but not of free radicals or atoms. As a result, this equation emphasizes the thermal mechanisms, and it has been widely used in correlations of experimental flame velocities.

The Semenov equation assumes an ignition temperature, but by approximations this temperature is eliminated from the final equation and thus the equation is made more useful.

The theory was advanced further when it was postulated that not only can heat control the reaction mechanism but possibly also the diffusion of certain active species such as radicals. It is quite possible that certain light particles can readily diffuse back and initiate further reactions.

This theory of particle diffusion was first put forth in 1934 by Lewis and von Elbe in dealing with the ozone reaction. Tanford and Pease (1947) postulated that it is the diffusion of radicals that is all important and not the temperature gradient as required by the thermal theories. They proposed a diffusion theory that was quite different in physical concept than the thermal theory. However, one should recall that equations which govern mass diffusion are the same as those which govern thermal diffusion.

After these theories were put forth, there was a great deal of experimentation in order to determine the effect of temperature and pressure on the flame velocity and thus to possibly check which theories were correct. In the thermal theory, the higher the ambient temperature the higher the final temperature and therefore the higher the reaction rate and flame velocity. Similarly for the diffusion theory, the higher the temperature, the higher the dissociation, the greater the concentration of radicals to diffuse back, and therefore the higher the velocity. Consequently, data obtained from temperature and pressure effects did not give conclusive results.

There appeared to be some evidence to support the diffusion concept, for this theory seemed to best explain the effect of H_2O on the experimental flame velocities of CO–O_2. At high temperature it is known that water provides the source of hydroxyl radicals to facilitate rapid reaction, as has been explained in previous sections.

Hirschfelder reasoned that in the cyanogen–oxygen flame there is no dissociation. The products of this reaction are CO and N_2, no intermediate species form and C=O and N≡N bonds are difficult to break. In this system it is apparent that the concentration of radicals is not important for flame propagation and one must conclude that thermal effects predominate. Hirschfelder *et al.* (1954) basically concluded that one should follow the thermal theory idea and include the diffusion of all particles both into and out of the flame zone. In setting up the equations governing the thermal and diffusional processes, Hirschfelder obtained a set of complicated nonlinear equations which could only be solved by numerical methods. In order to solve the set of equations, Hirschfelder had to assume some heat sink for a boundary condition on the cold side. This sink was required because of the use of the Arrhenius expressions for reaction rate. The complexity is that the Arrhenius expression requires a finite reaction rate even at $\chi = -\infty$, where the temperature is the unburned gas temperature.

Friedman and Burke (1953) in order to simplify the Hirschfelder solution modified the Arrhenius reaction rate equation so the rate was zero at $T = T_u$. But Friedman and Burke still required numerical calculations.

It was possible by certain physical principles to simplify the complete equations so that they could be solved relatively easily. The simplification was first carried out by von Karman and Penner (1954). This approach was considered one of the most significant advances in laminar flame propagation, but it could not have been developed and verified if it were not for the extensive work of Hirschfelder and his collaborators. The major simplification that von Karman and Penner introduced is the fact that the eigenvalue solution of the equations was the same for all ignition temperatures whether it be near T_f or not.

It is easily recognized that any exact solution of laminar flame propagation must make use of the basic equations of fluid dynamics modified to account for the liberation and conduction of heat and for changes of chemical species within the reaction zones.

The equations can be simplified by certain assumptions. Such assumptions have led to various other theories, but the theories that will be considered here are an extended development of Mallard–Le Chatelier and the Semenov approach. The Mallard–Le Chatelier approach is given because of its historical significance and because its very simple thermal approach permits the establishment of the important parameters in laminar flame propagation that are difficult to extract from the more exact theories. The Zeldovich–Frank-Kamenetskii–Semenov theory is reviewed because certain approximations used to obtain solutions are useful in problems arising in other areas of the combustion field.

1. The Theory of Mallard and Le Chatelier

Conceptually, Mallard and Le Chatelier stated that the heat conducted from zone II in Fig. 5 equalled that necessary to bring the unburned gases to the ignition temperature (the boundary between zone I and II). If it is assumed that the slope of the temperature curve was linear, then it could be approximated by the expression $[(T_f - T_i)/\delta]$, where δ is the thickness of the reaction zone. The enthalpy balance then becomes

$$\dot{m} c_p (T_i - T_0) = \lambda (T_f - T_i)/\delta \tag{8}$$

where λ is the thermal conductivity and $\dot{m}$ is the mass rate into the combustion wave. Since the problem as described is fundamentally one-dimensional,

$$\dot{m} = \rho A u = \rho S_L A \tag{9}$$

where ρ is the density, A the cross-sectional area taken as unity, u the velocity of the gases, and S_L the laminar flame speed. Because unburned gases enter normal to the wave, by definition

$$S_L = u \tag{10}$$

Equation (8) then becomes

$$\rho S_L c_p(T_i - T_0) = \lambda(T_f - T_i)/\delta \tag{11}$$

or

$$S_L = \frac{\lambda}{\rho c_p} \frac{(T_f - T_i)}{(T_i - T_0)} \frac{1}{\delta} \tag{12}$$

Equation (12) is the expression for the flame speed obtained by Mallard and Le Chatelier. In this expression δ is not known. Mallard and Le Chatelier really did not fully understand flame propagation and therefore a better representation is required.

It is possible to relate δ to S_L since δ is the reaction zone thickness. Thus

$$\delta = S_L \cdot \tau = S_L \frac{1}{d\varepsilon/dt} \tag{13}$$

where τ is the reaction time and $(d\varepsilon/dt)$ is the reaction rate in terms of fractional conversion as discussed in Section 2.D. Substituting this expression into Eq. (12), one obtains

$$S_L = \left(\frac{\lambda}{\rho c_p} \frac{T_f - T_0}{T_i - T_0} \frac{d\varepsilon}{dt}\right)^{1/2} \tag{14}$$

$$\sim (\alpha RR)^{1/2} \tag{15}$$

where $\alpha (= \lambda/\rho c_p)$ is the thermal diffusivity and RR is the reaction rate.

This adaptation of the simple Mallard–Le Chatelier approach is most significant in that the result

$$S_L \sim (\alpha RR)^{1/2}$$

is most powerful in predicting laminar flame spread phenomena. It is a result to be committed to memory. In fact one could question whether the more complex theories tell any more. Comparisons between actual calculated values of flame speed and experimental results are few and far between. The only published comparisons are for the simple hydrazine decomposition flame, and even there one could argue that the agreement is fortuitous. In any of the theories, in order to calculate the laminar flame speed, one must know the thermophysical properties of a complex mixture at high temperatures and have accurate reaction rate data. These types of data are just not available. The worth of the more advanced theories rests solely in the fact that they can obtain a quantitative result that compares within an order of magnitude to the experimental flame speed. In this sense they give validation to all the physical models based on heat and mass diffusion.

To examine how Eq. (15) predicts experimental trends, first recall that

$$d\varepsilon/dt = k\varepsilon^n p^{n-1} = Ae^{-E/RT}\varepsilon^n p^{n-1} \tag{16}$$

where n is the overall reaction order. Thus, it is possible to calculate the pressure dependency. Since for gases, only the density term in α is pressure dependent and

$$S_L \sim [(1/p)p^{n-1}]^{1/2} \sim (p^{n-2})^{1/2} \tag{17}$$

which states that the flame speed is independent of pressure for second-order reactions. Most hydrocarbon oxygen reactions have an overall reaction order close to 2 and thus Eq. (17) predicts that the flame speeds for hydrocarbons should be pressure independent. This pressure independence has been verified experimentally.

Even though it is not possible to evaluate T_i, the temperature dependence in the flame speed expression is dominated by the exponential; thus it is safe to assume

$$S_L \sim (e^{-E/RT})^{1/2} \tag{18}$$

Since physical reasoning states that most of the reaction must occur close to the highest temperature if Arrhenius kinetics control, the temperature to be used in the above expression is T_f and

$$S_L \sim (e^{-E/RT_f})^{1/2} \tag{19}$$

Thus the effect of varying the initial temperature is found in the degree it alters the flame temperature. Recall a hundred degree rise in initial temperature results in a rise of flame temperature which is much smaller. These specific trends due to temperature have been verified by experiment.

2. The Theory of Zeldovich, Frank-Kamenetskii, and Semenov

The Russian investigators derived expressions for the laminar flame speed by an important extension of the very simplified Mallard–Le Chatelier approach. Their basic equation included diffusion of species as well as heat.

Since their initial insight was that flame propagation was fundamentally a thermal mechanism, they were not concerned with the diffusion of radicals and their effect on the reaction rate, but rather with the energy transported by the diffusion of species.

As in the Mallard–Le Chatelier approach, an ignition temperature arises in this development; however, it is used only as a mathematical convenience for computation. Because the chemical reaction rate is an exponential function of temperature according to the Arrhenius equation, Semenov assumes that the ignition temperature, above which nearly all reaction occurs, is very near the flame temperature. Under these conditions the ignition temperature can be eliminated in the mathematical development.

For the initial development, two other important assumptions are made, although these restrictions are partially removed in further developments. The assumptions are that c_p and λ are constant and that

$$(\lambda/c_p) = D\rho \tag{20}$$

where D is the mass diffusivity. This assumption is essentially that $\alpha = D$. Developments in simple kinetic theory show that

$$\alpha = D = \nu \tag{21}$$

where ν = kinematic viscosity (momentum diffusivity). The ratios of these three diffusivities give some of the familiar dimensionless similarity parameters

$$\mathrm{Pr} = \nu/\alpha, \qquad \mathrm{Sc} = \nu/D, \qquad \mathrm{Le} = \alpha/D \tag{22}$$

where Pr, Sc, and Le are the Prandtl, Schmidt, and Lewis numbers, respectively. Thus from kinetic theory as a first approximation,

$$\mathrm{Pr} = \mathrm{Sc} = \mathrm{Le} = 1 \tag{23}$$

In addition, it can be shown that

$$(\lambda/c_p) = D\rho \neq f(P) \tag{24}$$

Consider the thermal wave given in Fig. 5. If a differential control volume is taken within this one-dimensional wave and the variations as given in the figure are in the x-direction, then the thermal and mass balances are as shown in Fig. 6. In Fig. 6, a is the number of moles of reactant per cubic centimeter, $\dot{w}$ is the rate of reaction, and Q is the heat of reaction per mole of reactant. Since the problem is a steady one, there can be no accumulation of species or heat with respect to time and the balance of the energy terms and the species terms must each be equal to zero.

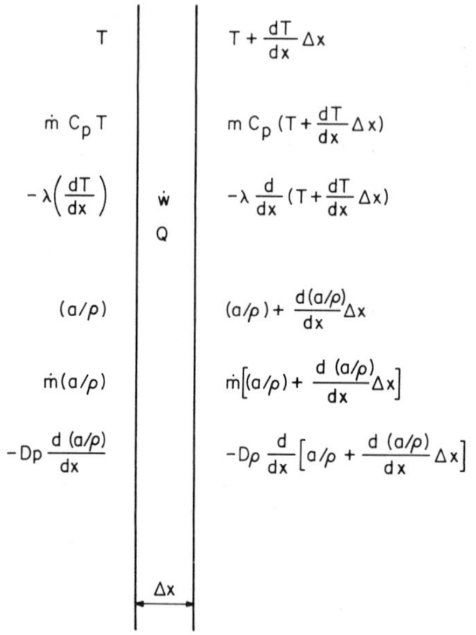

T | $T + \frac{dT}{dx}\Delta x$

$\dot{m}\, C_p T$ | $m\, C_p \left(T + \frac{dT}{dx}\Delta x\right)$

$-\lambda\left(\frac{dT}{dx}\right)$ | $\dot{w}$ Q | $-\lambda \frac{d}{dx}\left(T + \frac{dT}{dx}\Delta x\right)$

(a/ρ) | $(a/\rho) + \frac{d(a/\rho)}{dx}\Delta x$

$\dot{m}(a/\rho)$ | $\dot{m}\left[(a/\rho) + \frac{d\,(a/\rho)}{dx}\Delta x\right]$

$-D\rho \frac{d\,(a/\rho)}{dx}$ | $-D\rho \frac{d}{dx}\left[a/\rho + \frac{d\,(a/\rho)}{dx}\Delta x\right]$

Δx

Fig. 6. Balances across a differential element in a thermal wave.

The amount of mass convected into the volume $A\,\Delta x$ (where A is the area usually taken as unity) is

$$\dot{m}\left[(a/\rho) + \frac{d(a/\rho)}{dx}\Delta x\right]A - \dot{m}(a/\rho)A = \dot{m}\frac{d(a/\rho)}{dx}A\,\Delta x \tag{25}$$

The amount of mass diffusing into the volume is

$$\frac{d}{dx}\left[D\rho\left(\frac{a}{\rho} + \frac{d(a/\rho)}{dx}\Delta x\right)\right]A - \left[-D\rho\frac{d(a/\rho)}{dx}\right]A = -D\rho\frac{d^2(a/\rho)}{dx^2}A\,\Delta x \tag{26}$$

The amount of mass reacting (disappearing) in the volume is

$$-\dot{w}A\,\Delta x \tag{27}$$

Thus the continuity equation for the reactant is

$$\underset{\text{(diffusion term)}}{D\rho\frac{d^2(a/\rho)}{dx^2}} - \underset{\text{(convective term)}}{\dot{m}\frac{d(a/\rho)}{dx}} + \underset{\text{(generation term)}}{\dot{w}} = 0 \tag{28}$$

The energy equation is determined similarly and is

$$\lambda\frac{d^2T}{dx^2} - \dot{m}c_p\frac{dT}{dx} + \dot{w}Q = 0 \tag{29}$$

These equations are correct only if λ, c_p, and $D\rho$ do not depend on x, i.e., the temperature and composition. The state equation is written as

$$\rho/\rho_0 = (T_0/T) \tag{30}$$

New variables are defined as

$$\theta = c_p(T - T_0)/Q \tag{31}$$

$$\alpha = (a_0/\rho_0) - a/\rho \tag{32}$$

where the subscript 0 designates initial conditions. Substituting the new variables in Eqs. (28) and (29), two new equations are obtained

$$D\rho\, d^2\alpha/dx^2 - \dot{m}\, d\alpha/dx + \dot{w} = 0 \tag{33}$$

and

$$(\lambda/c_p)(d^2\theta/dx^2) - \dot{m}\, d\theta/dx + \dot{w} = 0 \tag{34}$$

The boundary conditions for these equations are

$$x = -\infty,\ \alpha = 0,\quad \theta = 0; \qquad x = +\infty,\quad \alpha = a_0/\rho_0,\quad \theta = [c_p(T_f - T_0)]/Q \tag{35}$$

where T_f is the final or flame temperature. For the condition $D\rho = (\lambda/c_p)$, Eqs. (33) and (34) are identical in form. If the equations and boundary conditions for α and θ coincide, i.e. if $\alpha = \theta$ over the entire interval, then

$$c_p T + aQ/\rho = c_p T_0 + a_0 Q/\rho_0 = c_p T_f \tag{36}$$

The meaning of Eq. (36) is that the sum of the thermal and chemical energies per unit mass of the mixture is constant in the combustion zone; i.e., the relation between the temperature and the composition of the gas mixture is the same as that for the adiabatic behavior of the reaction at constant pressure.

Thus the variable defined in Eq. (36) can be used to develop a new equation in the same manner as Eq. (28) and the problem reduces to the solution of only one differential equation. Indeed either Eq. (28) or (29) can be solved; however, Semenov dealt with the energy equation.

In the first approach it is assumed also that the reaction proceeds by zero order. Since the rate term $\dot{w}$ is not a function of concentration, the continuity equation is not required and it is possible to deal with the more convenient energy equation. The Russian investigators similar to Mallard–Le Chatelier examine the thermal wave as if it were made up of two parts. The unburned gas part is considered to be a zone of no chemical reaction, and the reaction part is considered to be the zone in which the reaction and diffusion terms dominate and the convective term can be dropped. Thus in zone I, the energy equation reduces to

$$d^2T/dx^2 - (\dot{m}c_p/\lambda)(dT/dx) = 0 \tag{37}$$

with the boundary conditions

$$x = -\infty, \quad T = T_0; \qquad x = 0, \quad T = T_i$$

It is apparent from the latter boundary condition that the coordinate system is so chosen that T_i is at the origin. The reaction zone extends a small distance d, so that in reaction zone II, the energy equation is written as

$$(d^2T/dx^2) + (\dot{w}Q/\lambda) = 0 \tag{38}$$

with the boundary conditions

$$x = 0, \quad T = T_i; \qquad x = d, \quad T = T_f$$

The added condition which permits the determination of the solution (eigenvalue) is the requirement of the continuity of heat flow at the interface of the two zones

$$\lambda(dT/dx)_{x=0,\,\mathrm{I}} = \lambda(dT/dx)_{x=0,\,\mathrm{II}} \tag{39}$$

The solution to the problem is obtained by initially considering Eq. (38). First, recall that

$$(d/dx)(dT/dx)^2 = 2(dT/dx)(d^2T/dx^2) \tag{40}$$

Now, multiply Eq. (38) by $2(dT/dx)$ and obtain

$$2(dT/dx)(d^2T/dx^2) = -2(\dot{w}Q/\lambda)(dT/dx) \tag{41}$$

which can be written as

$$(d/dx)(dT/dx)^2 = -2(\dot{w}Q/\lambda)(dT/dx) \tag{42}$$

Integrating Eq. (42), one obtains

$$-(dT/dx)^2_{x=0} = -2(Q/\lambda)\int_{T_i}^{T_f} \dot{w}\, dT \tag{43}$$

since $(dT/dx)^2$ evaluated at $x = d$ or $T = T_f$ is equal to zero. But from Eq. (37), one has

$$(d/dx)(dT/dx) = (\dot{m}c_p/\lambda)(dT/dx) \tag{44}$$

Integrating Eq. (44), one obtains

$$dT/dx = (\dot{m}c_p/\lambda)T + \text{const}$$

Since at $x = -\infty$, $T = T_0$ and $dT/dx = 0$,

$$\text{const} = -(\dot{m}c_p/\lambda)T_0 \tag{45}$$

and

$$dT/dx = \dot{m}c_p(T - T_0)/\lambda \tag{46}$$

Evaluating the expression at $x = 0$ where $T = T_i$, one obtains

$$(dT/dx)_{x=0} = \dot{m}c_p(T_i - T_0)/\lambda \tag{47}$$

The continuity of heat flux permits this expression to be combined with Eq. (43) to obtain

$$\dot{m}c_p(T_i - T_0)/\lambda = \left((2Q/\lambda)\int_{T_i}^{T_f} \dot{w}\, dT\right)^{1/2}$$

Since Arrhenius kinetics dominate, it is apparent that T_i is very close to T_f, so the last expression is rewritten as

$$\dot{m}c_p(T_f - T_0)/\lambda = \left((2Q/\lambda)\int_{T_i}^{T_f} \dot{w}\, dT\right)^{1/2} \tag{48}$$

Since $\dot{m} = S_L\rho_0$ and $(a_0/\rho_0)Q = c_p(T_f - T_0)$ (from Eq. (36)),

$$S_L = (2(\lambda/\rho c_p)\{I/(T_f - T_0)\})^{1/2} \tag{49}$$

where

$$I = (1/a_0)\int_{T_i}^{T_f} \dot{w}\, dT \tag{50}$$

Since $\dot{w}$ is a function of T and not of concentration for a zero order reaction, it may be written as

$$\dot{w} = Z' \exp(-E/RT) \tag{51}$$

where Z' is the preexponential term in the Arrhenius expression.

For sufficiently large energy of activation such as one has for hydrocarbon–oxygen mixtures where $E = O(40\ \text{kcal/mol})$, $(E/RT) \gg 1$. Thus most of the reaction will take place near the flame temperature and indeed T_i will be very near the flame temperature. Thus zone II is a very narrow zone. Consequently it is possible to define a new variable σ such that

$$\sigma = (T_f - T) \tag{52}$$

The values of σ will vary from

$$\sigma_i = (T_f - T_i) \tag{53}$$

to zero. Since $\sigma \ll T_f$, then

$$\exp[-E/RT] = \exp[-E/R(T_f - \sigma)] = \exp[-E/RT_f(1 - \sigma/T_f)]$$
$$= \exp[-(E/RT_f)(1 + \sigma/T_f)] = \exp[-E/RT_f]\exp[-E\sigma/RT_f^2]$$

Thus the integral I becomes

$$I = \frac{Z' \exp[-E/RT_f]}{a_0} \int_{T_i}^{T_f} \exp(-E\sigma/RT_f^2)\, dT$$
$$= -\frac{Z' \exp[-E/RT_f]}{a_0} \int_{\sigma_i}^{0} \exp[-E\sigma/RT_f^2]\, d\sigma \tag{54}$$

Defining still another variable β as

$$\beta = E\sigma/RT_f^2 \tag{55}$$

the integral becomes

$$I = (Z' e^{-E/RT_f}/a_0) \left[\int_0^{\beta_i} e^{-\beta}\, d\beta\right] (RT_f^2/E) \tag{56}$$

With sufficient accuracy one may write

$$j = \int_0^{\beta_i} e^{-\beta}\, d\beta = 1 - e^{-\beta_i} \cong 1 \tag{57}$$

since $(E/RT_f) \gg 1$ and $(\sigma_i/T_f) \cong 0.25$. Thus

$$I = (Z'/a_0)(RT_f^2/E)\, e^{-E/RT_f} \tag{58}$$

and

$$S_L = \left(\frac{2\lambda}{c_p \rho_0 a_0} \frac{Z' e^{-E/RT_f}}{(T_f - T_0)} \frac{RT_f^2}{E}\right)^{1/2} \tag{59}$$

In the preceding development, it was assumed that the number of moles did not vary during reaction. This restriction can be removed to allow the number to change in the ratio (n_r/n_p), which is the number of moles of reactant to product. Further, the Lewis number equal to 1 restriction can be removed to allow

$$(\lambda/c_p)/D\rho = A/B$$

where A and B are constants. With these restrictions removed, then the result for a first-order reaction becomes

$$S_L = \left\{\frac{2\lambda_f (c_{p_f}) Z'}{\rho_0 \bar{c}_p^{\,2}} \left(\frac{T_0}{T_f}\right) \left(\frac{n_r}{n_p}\right) \left(\frac{A}{B}\right) \left(\frac{RT_f^2}{E}\right)^2 \frac{e^{-E/RT_f}}{(T_f - T_0)^2}\right\}^{1/2} \tag{60}$$

and for a second-order reaction

$$S_L = \left\{ \frac{2\lambda c_{p_f}^2 Z' a_0}{\rho_0 (\bar{c}_p)^3} \left(\frac{T_0}{T_f}\right)^2 \left(\frac{n_r}{n_p}\right)^2 \left(\frac{A}{B}\right)^2 \left(\frac{RT_f^2}{E}\right) \frac{e^{-E/RT_f}}{(T_f - T_0)^3} \right\}^{1/2} \tag{61}$$

c_{p_f} is the specific heat evaluated at T_f and $\bar{c}_p$ is the average specific heat between T_0 and T_f.

Notice that since a_0 and ρ_0 are both functions of pressure, S_L is independent of pressure. Further this more complex development shows that

$$S_L \sim \left(\frac{\lambda c_{p_f}^2}{\rho_0 (\bar{c}_p)^3} a_0 Z\, e^{-E/RT_f} \right)^{1/2} \sim \left(\frac{\lambda}{\rho_0 c_p} RR \right)^{1/2} \sim (\alpha RR)^{1/2}$$

as was obtained from the simple Mallard–Le Chatelier approach.

3. The Laminar Flame and the Energy Equation

An important point about laminar flame propagation not previously discussed is worth stressing. It has become the practice to state that in the case of premixed homogeneous combustible gaseous mixtures that reaction rate phenomena control and in initially unmixed fuel-oxidizer systems that diffusion phenomena control. The subject of diffusion flames will be discussed in Chapter 6. In the case of laminar flames, and indeed in most aspects of turbulent flame propagation, it should be emphasized that it is the diffusion of heat (and mass) that causes the flame to propagate, i.e., flame propagation is a diffusional mechanism. The reaction rate determines the temperature gradient by its effect on the thickness of the reaction zone. The effect indeed is a strong one; nevertheless, there is a diffusional effect as well. The expression $S_L \sim (\alpha RR)^{1/2}$ says it well—the propagation rate is proportional to the square root of the diffusivity and the reaction rate.

4. Flame Speed Measurements

For many years there was no interest in flame speed measurements. Sufficient data and understanding were thought to be at hand. As lean burn conditions became popular in spark ignition engines, interest in flame speeds of lean limits has rekindled this interest in measurement techniques. Some techniques are discussed in the paragraphs which are to follow.

The flame velocity has been defined as the velocity at which the unburned gases move through the combustion wave in a direction normal to the wave surface. If, in an infinite plane flame, the flame is regarded as stationary and a particular flow tube of gas considered, the area of the flame enclosed by the tube is not dependent on how the term "flame surface or wave surface" in which the area is measured is defined. The areas of all parallel surfaces are the same whatever property (e.g., particularly temperature) is chosen to define the surface, and these areas are all equal to each other and to that of the inner surface of the luminous part of the flame. The definition is more difficult in any other geometric system. Consider, for example, an experiment in which gas is supplied at the center of a sphere and flows radially outwards in a laminar manner to a spherical flame which is stationary. The inward movement of the flame is balanced by the outward flow of gas. The experiment takes place in an infinite volume at constant pressure. The area of the surface of the wave will depend on where the surface is located. The area of the sphere for which $T = 500°C$ will be less than that of one for which $T = 1500°C$. So, if the burning velocity is defined as the volume of unburned gas consumed per second divided by the surface area of the flame, the result obtained will depend on the particular surface selected. The only quantity that does remain constant in this system is $u_r \rho_r A_r$, where u_r is the velocity of flow at the radius r, where the surface area is A_r and the gas density is ρ_r. This product equals $\dot{m}$, the mass flowing through the layer at r per unit time, and must be constant for all values of r. Thus u_r varies with r, the distance from the center in the manner shown in Fig. 7.

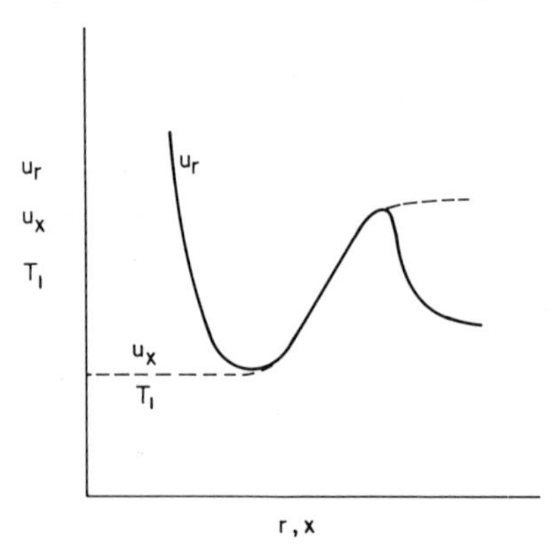

Fig. 7. Velocity and temperature variations through non-one-dimensional flame systems.

It is apparent from Fig. 7 that it is difficult to select a particular linear flowrate of unburned gas up to the flame and regard this velocity as the burning velocity.

If an attempt is made to define burning velocity for just such a system, it is found that no definition free from all possible objections can be formulated. Moreover, it is impossible to construct a definition which will, of necessity, lead to the same value being determined as that in an experiment using a planar flame. The essential difficulties are that over no range of r values does the linear velocity of the gas have even an approximately constant value and that, in this ideal system, the temperature varies continuously from the center of the sphere outwards and approaches the flame surface asymptotically as r approaches infinity. So no spherical surface can be considered to have a significance greater than any other.

In Fig. 7, u_x, the velocity of gas flow at x for a planar flame, is plotted on the same scale against x, the space coordinate measured normal to the flame front. It is assumed that over the main part of the rapid temperature rise u_r and u_x coincide. This is likely to be true if the curvature of the flame is large compared with the flame thickness. The burning velocity is then, strictly speaking, the value to which u_x approaches asymptotically as x approaches minus infinity. However, because the temperature of the unburned gas varies exponentially with x, the value of u_x becomes effectively constant only a very short distance from the flame. The value of u_r on the low temperature side of the spherical flame will not at any point be as small as the limiting value of u_x which one calls the burning velocity. However, the value of u_r at the point where it is a minimum is likely for all ordinary flames (for which the flame zone is thin) to be very little greater than the limiting value of u_x. In fact, the difference, though not zero, will probably be inappreciable for such flames. This value of u_r could be determined using the formula

$$u_r = \dot{m}/\rho_r A_r$$

Since the layer of interest is right on the unburned side of the flame, ρ_r will be close to ρ_u the density of the unburned gas and $\dot{m}/\rho_r$ will be close to the volume flowrate of unburned gas.

To obtain in practice a value for the burning velocity which is close to that for the planar flame it is necessary to locate and measure an area as far on the unburned side of the flame as possible. Systems such as Bunsen flames are in many ways more complicated than either the planar case or the spherical case.

Before proceeding, consider the methods of observation. The following methods have been most widely used to observe the flame:

A. The luminous part of the flame is observed and the side of this zone which is toward the unburned gas is used for measurement (direct photography).

B. A shadowgraph picture is taken.

C. A Schlieren picture is taken.

D. Interferometry (another less frequently used method).

Which surface in the flame does each method give? Again consider the temperature distribution through the flame as given in Fig. 8. The luminous zone comes late in the flame and thus direct photography is generally not a satisfactory method.

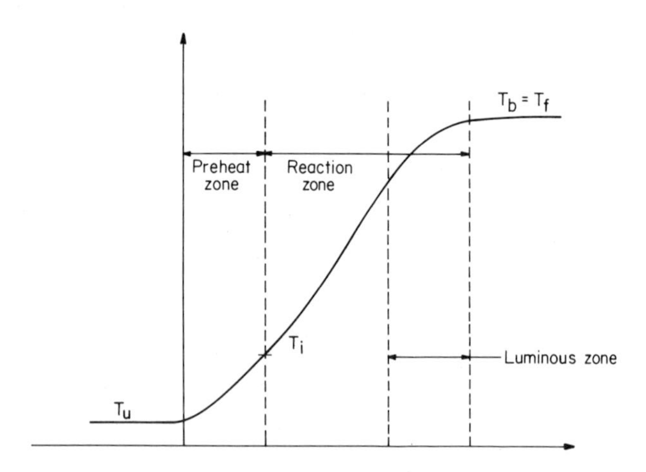

Fig. 8. Temperature regimes in a laminar flame.

Shadowgraph measures the derivative of the density gradient $(\partial\rho/\partial x)$ or $(-1/T^2)(\partial T/\partial x)$, i.e., it evaluates $\{\partial[(-1/T^2)(\partial T/\partial x)]/\partial x\} = (2/T^3)(\partial T/\partial x)^2 - (1/T^2)(\partial^2 T/\partial x^2)$. Shadowgraphs thus measure the earliest variational front and are not a precisely defined surface. Actually it is possible to define two shadowgraph surfaces, one at the unburned side and one on the burned side. The inner term is much brighter than the outer value since the absolute value for the expression above is greater when evaluated at T_u than at T_b.

Schlieren photography gives $(\partial\rho/\partial x)$ or $(-1/T^2)(\partial T/\partial x)$ which has the greatest value about the inflection point of the temperature curve and which corresponds more closely to the ignition temperature. This surface lies early in the flame, is more readily definable than most images, and is recommended and preferred by many workers. Interferometry, which measures density or temperature directly, is much too sensitive and can be used only on two-dimensional flames. In an exaggerated picture of a Bunsen tube the surfaces would lie as shown in Fig. 9.

The various experimental configurations used for flame speeds may be classified under the following headings:

(a) conical stationary flames on cylindrical tubes and nozzles;
(b) flames in tubes;

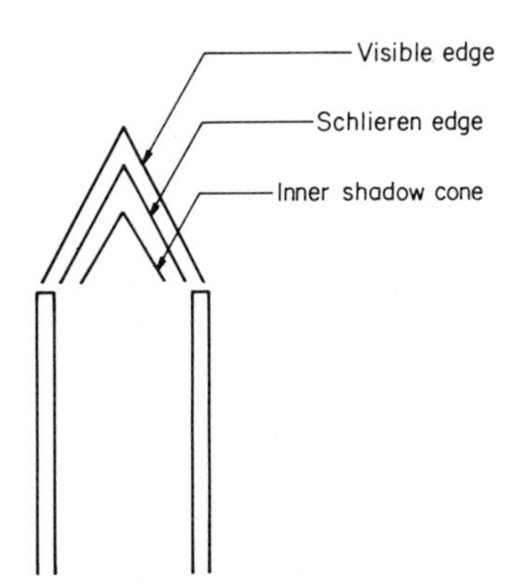

Fig. 9. Optical fronts in a Bunsen flame.

(c) soap bubble method;
(d) constant volume explosion in a spherical vessel;
(e) flat flame methods.

The methods are listed in order of decreasing complexity of flame surface and correspond to an increasing complexity of experimental arrangement. Each has certain advantages which bring about their usage.

a. Bunsen Burner Method

In this method premixed gases flow up a cylindrical jacketed tube long enough to insure streamline flow at the mouth. The gas burns at the mouth of the tube and the shape of the Bunsen cone is recorded and measured by various means and in various ways. When shaped nozzles are used instead of long tubes, the flow is uniform instead of parabolic and the cone has straight edges. Because of the complicated flame surface, the different procedures used for measuring the flame cone have led to different results.

The burning velocity is not constant over the cone. The velocity near the tube wall is lower because of cooling by the walls. Thus there are lower temperatures; therefore, lower reaction rates; consequently, lower flame speeds. The top of the cone is crowded due to the presence of large energy release and therefore reaction rates are too high.

It has been found that 30% of the internal portion of the cone gives a constant flame speed when related to the proper velocity vector and thus gives results comparable with other methods.

Actually, if one measures S_L at each point he will see that it varies along every point for each velocity vector and it is not really constant. This is the major disadvantage of this method.

The earliest procedure of calculating flame speed by this method was to divide the volume flowrate by the area of flame cone.

$$S_L = \frac{V}{A} \frac{\text{cm}^3/\text{sec}}{\text{cm}^2} = \text{cm/sec}$$

It is apparent then that the choice of cone will give widely different results. Experiments with fine magnesium oxide particles dispersed in the gas stream have shown that the flow streamlines remain relatively unaffected until the Schlieren cone and then diverge from the Bunsen axis before reaching the visible cone. These experiments have led many investigators to use the Schlieren cone as the most suitable one for flame speed evaluation.

The shadow cone is used by many experimenters because of the much greater simplicity than Schlieren techniques. The shadow being on the cooler side certainly gives more correct results than the visible cone. The fact that the flame cone can act as a lens in shadow measurements causes uncertainties as to the proper cone size.

Some investigators have used the central portion of the cone only and used the volume flow through tube radii corresponding to this portion. The proper choice of cone is of concern here also.

The angle the cone slant makes with the burner axis can also be used to determine S_L (see Fig. 10). This angle should be measured only at the central portion of the cone. Thus $S_L = u_u \sin \alpha$.

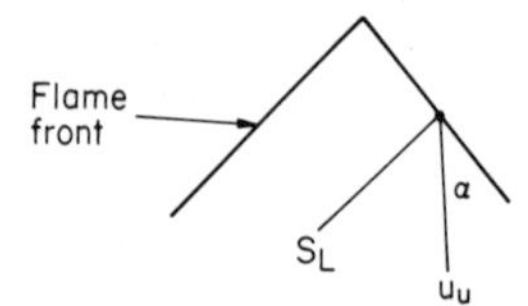

Fig. 10. Velocity vectors in a Bunsen cone.

Two of the disadvantages of the Bunsen method are

(1) One can never completely eliminate wall effects.

(2) One needs a steady source of supply of gas, which for rare or pure gases can be a severe problem.

The next three methods to be discussed make use of small amounts of gas.

b. Cylindrical Tube Method

A gas mixture in a horizontal tube opened at one end is ignited at the open end. The rate of progress of the flame into the unburned gas is the flame speed.

The difficulty with this method is that the flame front is curved due to buoyancy effects. Then the question arises as to which area in the flame is to be used. The flame area is no longer the geometric image of the tube; if hemispherical, $S_L A_f = u_m \pi R^2$. Closer observation will also reveal quenching at the wall and mixing of the unaffected center with an affected peripheral area.

The gas ahead of the flame is affected by the flame because a pressure wave is established by the burning (heating). This pressure wave causes a velocity in the unburned gases and it is necessary to account for this gas movement. This velocity must be subtracted from the measured value since the flame is propagating within a moving gas. Friction effects downstream cause a greater pressure wave, therefore length can have an effect. One can put a hole at the end of the tube, drill a small hole in the cap and measure the efflux with a soap solution (Gerstein *et al.*, 1951). The growth of the soap bubble is used to obtain the velocity of gases leaving the tube and the velocity of the unburned gas. Another physical restriction at the open end minimizes effects due to the back flow of expanded gases.

These adjustments permit relatively good values to be obtained, but still there are errors due to wall effects and distortion effects due to buoyancy. This buoyancy effect is modified by turning the tube vertically.

c. *Soap Bubble Method*

In an effort to eliminate wall effects, two spherical methods were developed. In the one to be discussed, the gas mixture is contained in a soap bubble and ignited at the center by a spark so that a spherical flame spreads radially through the gas. Because the gas is enclosed in a soap film, the pressure remains constant. The growth of the flame front along a diameter is followed by some photographic means. Because, at any stage of the explosion, the burned gas behind the flame occupies a larger volume than it did as unburned gas, the fresh gas into which the flame is burning moves outwards. Then

$$S_L \cdot A \cdot \rho_u = u_r \cdot A \cdot \rho_b$$

amount of material which must go into flame to increase volume = observed mass increase

$$S_L = u_r(\rho_b/\rho_u)$$

where u_r is the observed flame velocity.

The major disadvantage of this method is the large uncertainty in the temperature ratio T_u/T_b necessary to obtain ρ_b/ρ_u. Other disadvantages are

(1) The method can only be used for fast flames to avoid the convective effect of hot gases.

(2) The method cannot work with dry mixtures.

d. Closed Spherical Bomb Method

The bomb method is quite similar to the bubble method except the constant volume condition causes a variation in pressure. The pressure must be followed simultaneously with the flame front.

Similar to the soap bubble method, only fast flames can be used because the adiabatic compression of the unburned gases must be measured in order to calculate the flame speed. Also, the gas into which the flame is moving is always changing and consequently both the burning velocity and flame speed vary through the explosion. These features make the treatment complicated and, to a considerable extent, uncertain. The following expression has been derived (Fiock *et al.*, 1940) for the flame speed:

$$S_{\mathrm{L}} = [1 - [(R^3 - r^3)/3p\gamma_{\mathrm{u}} r^2](dp/dr)](dr/dt)$$

where R is the sphere radius, r the radius of spherical flames at any moment, and γ_{u} the ratio of specific heats of the unburned gases.

The fact that the second term in the brackets is close to 1 makes it difficult to attain high accuracy.

e. Flat Flame Burner Method

The flat flame burner method is usually attributed to Powling (1949). It is probably the most accurate because it offers the most simple flame front and one in which the area of the shadow, Schlieren, and visible fronts are all the same.

By placing a porous metal disk or a series of small tubes of 1-mm diameter or less at the exit of the larger flow tube, it is possible to create suitable conditions for flat flames. The flame is usually ignited with too high a flowrate and the flow or composition adjusted until the flame is flat. The diameter of the flame is then measured and the area divided into the volume flowrate of unburned gas. If the velocity emerging is greater than the flame speed, a larger flame area is required and a cone is obtained. If velocity is too slow, the flame tends to flash back and is quenched. In order to accurately define the edges of the flame, an inert gas is usually flowed around the burners. By controlling the rate of efflux of burned gases by a grid, a more stable flame is obtained. This experimental apparatus would appear as shown in Fig. 11.

This method was applicable only to mixtures having low burning velocities of the order of 15 cm/sec and less. At higher burning velocities,

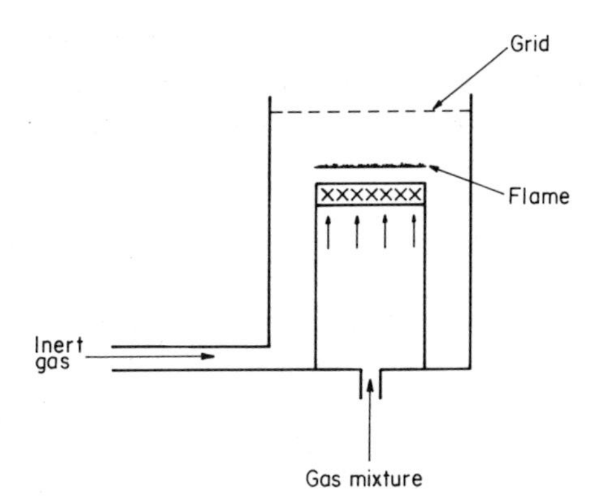

Fig. 11. A flat flame apparatus.

the flame front positions itself is too far from the burner and forms conical shapes.

Spalding and Botha (1954) ingeniously extended the flame burner method to higher flame speeds by cooling the plug. The cooling brings the flame front closer to the plug and stabilizes it. Operationally, the procedure is as follows: A flowrate giving a velocity greater than the flame speed is set and the cooling controlled until a flat flame is obtained. For a given mixture ratio many cooling rates are used. A plot of flame speed versus cooling rate is made and is extrapolated to zero cooling rate (Fig. 12). At

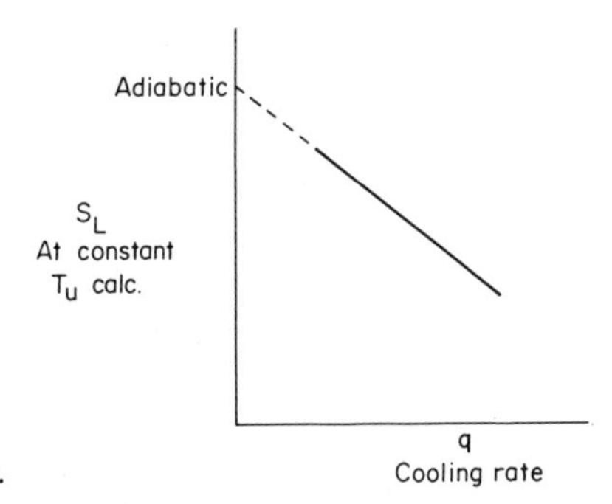

Fig. 12. The cooling effect in flat flame burners.

this point the adiabatic flame speed S_L is obtained. This procedure can be used for all mixture ratios within the inflammability limits.

The reason this procedure is preferable over the other methods is that the heat which is generated is leaking to the porous plug and not to the unburned gases as in the other model. Thus there is quenching all along the plug and not just at the walls.

The temperature at which the flame speed is measured is calculated as follows. For the approach gas temperature, one calculates what the initial temperature would have been if there were no heat extraction. Then the velocity of the mixture which would give the measured mass flowrate at this temperature is calculated. This velocity is S_L at the calculated temperature.

5. Experimental Results and Physical and Chemical Effects

Consideration has been given already to physical effects when the results of the Mallard–Le Chatelier approach were discussed. There are two important considerations—the thermal diffusivity and the temperature. The pressure effect is straightforward; the flame speed varies as $P^{(n-2)/2}$. A very extensive compilation of early flame speed measurements is given in NACA Rep. 1300.

Certain other points are worth discussing, however. The variation of flame speed with fuel-oxidant ratio follows the variation of temperature with the mixture ratio. Only in hydrogen mixtures does the thermal diffusivity play a role. Thus the flame speed will peak at the stoichiometric mixture ratio. Most hydrocarbons have the same stoichiometric flame temperature for burning in oxygen systems; consequently, most hydrocarbons will have the same flame speeds. Attempts have been made to correlate the flame speed with the fuel structure (NACA Rep. 1300). When such correlations have been found, it also becomes apparent that the flame temperature can be correlated with fuel structure, and indeed the temperature is the significant variable. Typical results of flame speeds as a function of fuel–air ratio are shown in Fig. 13.

The effect of the initial temperature on the flame propagation rate again appears to be reflected upon the effect on the final or flame tempera-

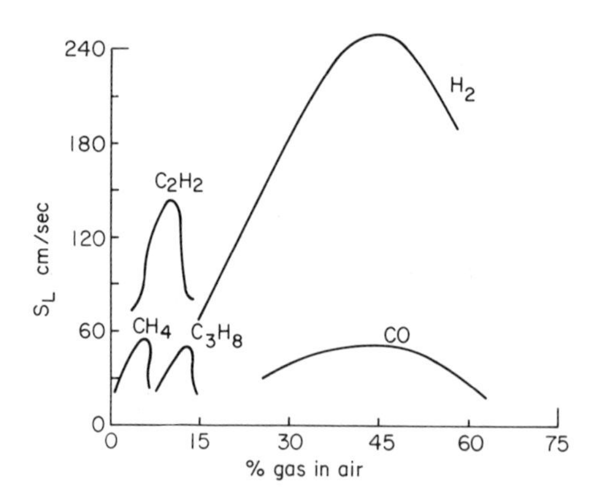

Fig. 13. The variation of flame speed with mixture ratio.

ture. Generally, small changes of initial temperature have little effect on the flame temperature since the chemical energy release is always so much greater than the sensible energy of the reactants. Nevertheless, the flame propagation expression contains the flame temperature in an exponential term; thus small changes in flame temperature can give noticeable changes in flame propagation rates.

Perhaps the most interesting set of experiments to elucidate the dominant factors in flame propagation was performed by Clingman *et al.* (1953). Their results certainly clearly show the effects of the thermal diffusivity and reaction rate terms. These investigators measured the flame propagation rate of methane in various oxygen–inert gas mixtures. The mixture of oxygen to inert gas was always 0.21/0.79 on a volumetric basis, the same as that which exists for air. The inerts chosen were nitrogen (N_2), helium (He), and argon (Ar). The results of these experiments are shown in Fig. 14.

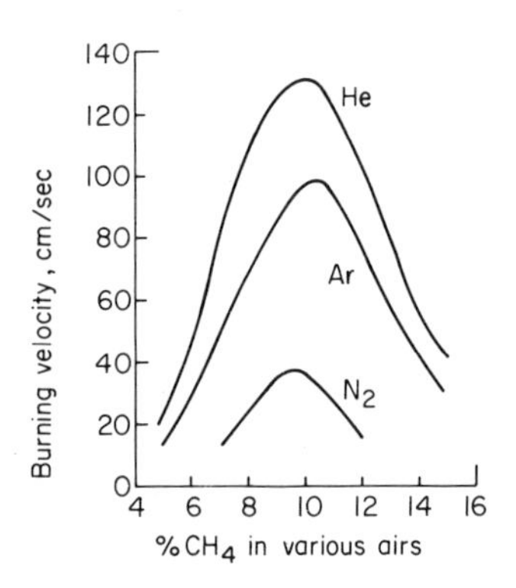

Fig. 14. Methane flame velocities in various airs (after Clingman et *al.*, 1953).

The trends of the results in Fig. 14 can be readily explained. Ar and N_2 have thermal diffusivities which are nearly equal. However, Ar is a monatomic gas which has a lower specific heat than N_2. Since the heat release in all systems is the same, the final (or flame) temperature will be higher in the Ar mixture than in the N_2 mixture. Thus S_L will be higher for Ar than N_2. Ar and He are both monatomic, and thus their final temperatures are equal. However, the thermal diffusivity of He is much greater than that of Ar. He has a higher thermal conductivity and a much lower density than Ar. Consequently, S_L for He is much greater than that for Ar.

The effect of chemical additives on the flame speed has also been explored extensively. Leason (1953) has reported the effects on flame velocity of small concentrations of additives ($<3\%$) and other fuels. He studied the propane–air flame and additives considered were compounds such as acetone, acetaldehyde, benzaldehyde, diethyl ether, benzene, and carbon disulfide, and many which were chosen from those classes of compounds which were oxidation intermediates in low-temperature studies and hence were expected to decrease the induction period and thus increase the flame velocity. Despite differences in apparent oxidation properties, all compounds changed the flame velocity in exactly the same way that dilution with excess fuel would on the basis of oxygen requirement. These results support the contention that the laminar flame speed is controlled by the high temperature reaction region. The high temperatures generate more than ample radicals via chain branching so that it is not likely that any additive could contribute any reaction rate accelerating feature.

There is, of course, a chemical effect in carbon monoxide flames. This point was mentioned in the discussion of carbon monoxide explosion limits. Studies have shown that CO flame velocities increase appreciably when small amounts of hydrogen, hydrogen containing fuels, or water are added. For 45% CO in air, the flame velocity passes through a maximum after approximately 5% by volume of water has been added. At this point the flame velocity is 2.1 times the value with 0.7% H_2O added. After 5%, a dilution effect begins to cause a decrease in flame speed. The effect and the maximum arise due to the necessity of establishing a sufficient steady-state concentration of OH radicals for the most effective explosive condition.

Although it may be expected that the common antiknock compounds would especially decrease the flame speed, no effects of antiknocks have been found in constant pressure combustion. The effect of the inhibition of the preignition reaction on flame speed is of negligible consequence. Although there is not universal agreement as to the mechanism of antiknock, many believe they serve to decrease the radical concentrations by particle surfaces. The reduction of the radical concentration in the preignition reactions or near the flammability limits can have severe consequences on the ability to initiate combustion. In these cases the radical concentrations are such that the chain branching factor is very close to the critical value for explosion. Any reduction could prevent the explosive condition from being reached. Around the stoichiometric mixture ratio the radical concentrations are normally so great that it would appear most difficult to add any small amounts of additives which would capture sufficient amounts of radicals to alter the reaction rate and the flame speed.

Certain halogen compounds such as the Freons are known to alter the flammability limits of hydrocarbon–air mixtures. The accepted mechanism is that the halogen atoms trap H radicals necessary for the chain-branching step. Near the flammability limits, α_{exp} is just above α_{crit} for explosion. Any reduction in radicals and the chain branching effects these radicals may have could eliminate the explosive (fast reaction rate and larger energy release rate) regime. However, small amounts of halogen compounds do not seem to affect the flame speed in a large region around the stoichiometric mixture ratio. The reason is again that in this region the temperatures are so high and radicals so abundant that elimination of some does not affect the reaction rate.

Recently it has been found that some of the longer chain halons (the generic name for the halogenated compounds sold under commercial names such as the Freons) are effective as flame suppressants. Also, some investigators have found that inert powders are effective in fire fighting. Fundamental experiments to evaluate the effectiveness of the halons and powders have been performed with an apparatus which measures the laminar flame speed. Results have been reported that the halons and the powders reduce flame speeds even around the stoichiometric air–fuel ratio. The investigators performing these experiments have argued that those agents are effective by reducing the radical concentrations. However, this explanation could be questioned. The quantities of these agents added are sufficient that they could absorb sufficient amounts of heat to reduce the temperature and thus the flame speed. Both halons and powders have large total heat capacities.

B. STABILITY LIMITS OF LAMINAR FLAMES

There are two types of limits generally associated with laminar flame propagation. One type governs the ability of the mixture to support flame propagation and the other is concerned with flow conditions, particularly in tubes. The first type includes flammability limits and the quenching distance and the second the phenomena of flashback, blowoff, and the onset of turbulence.

1. Flammability Limits

The explosion limit curves presented earlier were for a definite fuel–oxidizer mixture ratio. Indeed most were for the stoichiometric mixture ratio. In this case, even with a mixture at a very low temperature and reasonable pressures, if an ignition source were introduced into the mixture, the gases would move into the explosive region and a flame would propagate. The flame would propagate even after the ignition source was removed. There are mixture ratios, however, which will not self-support the flame after the ignition source is removed. These mixture ratios fall at the lean and rich end of the concentration spectrum. The leanest and richest concentrations which will just self-support a flame are called the lean and rich flammability limit, respectively. Flammability limits in both air and oxygen are found in the literature. The lean limit rarely differs for air or oxygen, as the excess oxygen in the lean condition has the same thermophysical properties as nitrogen.

a. Experimental Results and Physical and Chemical Effects

There have been attempts to standardize the determination of inflammability limits. Coward and Jones (1951) recommended that a 2-in. glass tube be employed, and it should be about 4 ft long and ignited by a spark a few millimeters long or by a small naked flame.

The high energy starting conditions are such that weak mixtures will be sure to ignite and the large tube diameter is selected because the most consistent results are obtained with tubes of this diameter. Quenching effects may interfere in tubes of small diameter. Large diameters create some disadvantages as the quantity of gas is a hazard and there is also the possibility of cool flames. The 4-ft length is to allow an observer to truly judge whether the flame will propagate indefinitely or not.

It is important to specify the direction of flame propagation. Since it may be stated as a rough approximation that a flame cannot propagate downward in a mixture if the convection current it produces is faster than the speed of the flame, the limits for upward propagation are usually slightly wider than for downward or horizontal propagation.

TABLE 1

Air flammability limits of some fuels

	Lower	Upper	Stoichiometric
Methane	5	14	9.47
Heptane	1	6	1.87
Hydrogen	4	74.2	29.2
Carbon monoxide	12.5	74.2	29.5
Acetyladehyde	3.97	57.0	7.7
Acetylene	2.50	80	7.7
CS_2	1.25	50	
Ethylene oxide	3.00	100	

Table 1 lists upper and lower flammability limits (in air) given by Smith and Stinson (1952).

In view of the accelerating effect of temperature on chemical reactions, it is reasonable to expect that limits of inflammability should be broadened if the temperature is increased. This trend is confirmed experimentally. The increase is only slight, and it appears to give a linear variation for hydrocarbons.

The behavior of the limits at elevated pressures has been explained somewhat satisfactorily. For simple hydrocarbons (ethane, propane, pentane, etc.), it appears that the rich limits extend almost linearly with increasing pressure at a rate of about 0.13 vol % per atm; the lean limits, on the other hand, are at first extended slightly and are thereafter narrowed as pressure is increased to 6 atm. In all, the lean limit is not affected appreciably by the pressure. Figure 15 for natural gas in air shows the effect of high pressure on flammability limits.

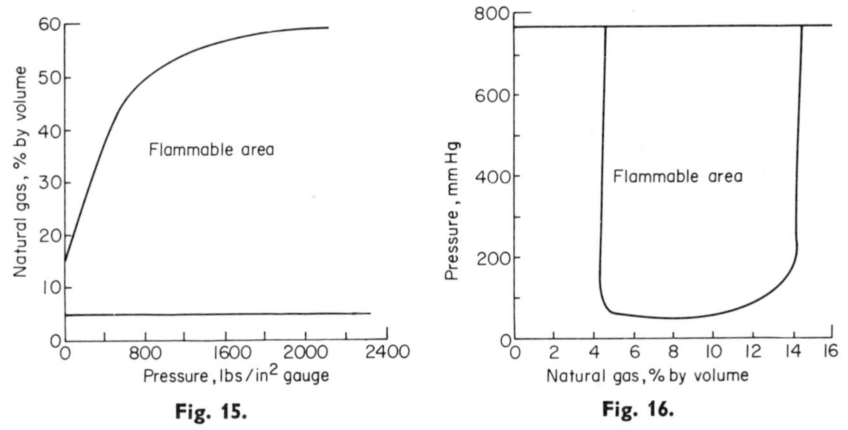

Fig. 15. **Fig. 16.**

Fig. 15. Effect of pressure increase above atmospheric on flammability limits of natural gas–air mixtures (after Jones *et al.*, 1949).

Fig. 16. Effect of reduction of pressure below atmospheric on flammability limits of natural gas–air mixtures (after Jones and Kennedy, 1945).

For flammability limits at reduced pressures, most of the older work indicated that the rich and lean limits converge as the pressure is reduced until a pressure is reached below which no flame will propagate. However, it was found that this behavior was due to wall quenching by the tube in which the experiments were performed. As shown in Fig. 16, the limits are actually as wide at low pressure as at 1 atm, provided the tube is sufficiently wide and provided an ignition source can be found to ignite the mixtures. Consequently the limits obtained at reduced pressures are not generally true limits of flammability since they are influenced by the tube diameter and are therefore not physicochemical constants of a given fuel. These low-pressure limits might better be termed limits of flame propagation in a tube of specified diameter.

A general "rule of thumb" is that the upper limit is about three times stoichiometric and the lower limit is about 50% of stoichiometric. Generally the upper limit is higher than that for detonation. The lower limit of a gas is the same in oxygen as in air. The higher limit of all flammable gases is much greater in oxygen than in air; hence, the range of flammability is always greater in oxygen. Table 2 shows this effect.

TABLE 2

Comparison of oxygen and air flammability limits

	Lower		Higher	
	Air	O_2	Air	O_2
H_2	4	4	74	94
CO	12	16	75	94
NH_3	15	15	28	79
CH_4	5	5	14	61
C_3H_8	2	2	10	55

As increasing amounts of an incombustible gas or vapor are added to the atmosphere, the flammability limits of a gaseous fuel in the atmosphere approach one another and finally meet. Inert diluents such as CO_2, N_2, or Ar merely replace part of the O_2 in the mixture, but they do not all have the same extinction power. It is found that the order of efficiency is the same as the order of the heat capacities of these three gases:

$$CO_2 > N_2 > Ar$$

For example, the minimum percent oxygen that will permit flame propagation in mixtures of CH_4, O_2, and CO_2 is 14.6%; if N_2 were the diluent, the minimum percent oxygen is less and equals 12.1%. In the case of Ar, the value is 9.8%. As discussed in many other chapters, a higher specific heat gas present in sufficient quantities will reduce the final temperature, and in this sense reduces the rate of energy release which must sustain the rate of propagation over other losses.

Other types of diluents are far more effective than the inert gases. In particular, halogenated hydrocarbons are most effective. For example, the effectiveness order compared to inert gases would be

$$CCl_4 > CO_2 > N_2 > Ar$$

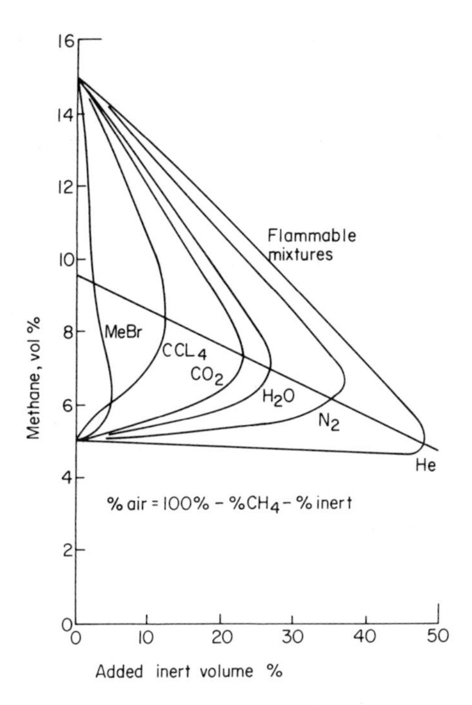

Fig. 17. Limits of flammability of various methane–inert gas–air mixtures at 25°C and 1 atm (after Zabatakis, 1965).

Even though carbon tetrachloride has a higher specific heat than the other gases, Fig. 17 taken from Zebatakis (1965) shows that its power is far more than one could expect from a temperature (thermal) argument alone. Thus it would appear that halogenated compounds affect the actual combustion mechanisms rather than the flame temperature alone. Again, it appears that the halogen atoms formed by decomposition attach to the H radicals to form the well-bonded hydrogen halide. This depletion of H radicals affects the chain branching. However, there is dispute concerning this point.

Notice in Fig. 17 that the rich limits are more sensitive to inert diluents than the lean limits; however, species which affect the reaction mechanism affect both limits. The Zebatakis report is the source of a wealth of data on flammability limits, flame speed, autoignition, etc.

b. A Theoretical Approach

The best theoretical approach for the prediction of flammability limit trends is due to Spalding (1957). This analysis is also extended to deal with quenching effects which are so intimately associated with the flammability limit problem.

Spalding's prediction of flammability is based upon a heat loss term, which arises due to a conduction or radiation loss or both. He considers the one-step reaction

$$A + B \longrightarrow C$$

in which B is considered to be in excess, so that the products downstream from a one-dimensional flame are B and C. Variations in stream properties normal to the x-direction are considered absent. Thus for heat losses one can visualize radiation being represented as a one-dimensional process; however, the heat conduction loss is an idealization of a two- or three-dimensional process.

The Semenov thermal theory is written with the Lewis number equal to 1, average specific heats, and a heat loss term as

$$\frac{d}{dx}\left(k\frac{dT}{dx}\right) - cG\frac{dT}{dx} + HR_A = L \qquad \text{(energy)} \qquad (62)$$

$$\frac{d}{dx}\left(\frac{k}{c}\frac{dm_A}{dx}\right) - G\frac{dm_A}{dx} + R_A = 0 \qquad \text{(species diffusion)} \qquad (63)$$

$$(m_{Bu} - m_B) = (m_{Au} - m_A)r \qquad \text{(stoichiometry)} \qquad (64)$$

where k is the thermal conductivity; G the total mass velocity per unit area; m the mass of the component per unit mass of mixture; R_A the consumption rate; H the standard heat of reaction of A; L the rate of heat transfer per unit volume; and r the stoichiometric ratio.

The equations are nondimensionalized by the substitutions

$$\tau = (T - T_u)/(T_b - T_u) \qquad (65)$$

$$\xi = \exp[cGx/k] \qquad (66)$$

$$\alpha = m_A/m_{Au} \qquad (67)$$

$$\lambda = HR_A{}^*k_b/(T_b - T_u)(cG)^2 \qquad (68)$$

$$\phi = kR_A/k_b R_A{}^* \qquad (69)$$

$$\psi = kL/k_b L^* \qquad (70)$$

$$K = L^*/R_A{}^*H \qquad (71)$$

$R_A{}^*$ is the value of R_A when $T = T_b$ and $m_A = m_{Au}$. This definition of a rate is a very unusual one since when $T = T_b$, $m_A = 0$ and not m_{Au}. L^* is the value of L at $T = T_b$. The subscripts u and b refer to the unburned and burned states, respectively. The particular choice for $R_A{}^*$ is necessary because the heat loss is always a power of T, and one will always want

$$\phi \sim \alpha\tau \tag{72}$$

which arises from

$$\phi \sim m_A T^m / m_{Au} T_b{}^m \tag{73}$$

The term ξ is an interesting substitution. In diffusion equations containing rate terms, ξ is a distributed space variable which eliminates the first derivative (convective) term.

Substituting the nondimensionalized variables in Eqs. (62) and (63), one obtains

$$d^2\tau/d\xi^2 = -(\lambda/\xi^2)\{\phi - K\psi\} \tag{74}$$

$$= \lambda/\xi^2 \, \phi \tag{75}$$

where

$$\phi = \phi(\alpha, \tau); \qquad \psi = \psi(\tau) \tag{76}$$

ϕ is a reaction term and ψ is a heat release term. The boundary conditions for the problem are

$$\begin{aligned} x = -\infty, \quad \xi = 0; \qquad \xi = 0, \quad \tau = 0, \quad \alpha = 1; \\ \xi = +\infty, \quad \xi = 0, \quad \alpha = 0 \end{aligned} \tag{77}$$

The $+\infty$ boundary conditions take the form given since the temperature profile must eventually become level and m_A must become zero. At $\xi = +\infty, \tau = 1$ is a mathematical inconvenience, and requires a new boundary condition to be written as

$$\xi = 1, \alpha = 0, \qquad (d\tau/d\xi) = -(d\alpha/d\xi) = (d\tau/d\xi)_1 \longrightarrow 0 \tag{78}$$

since for $K = 0$, $(\alpha + \tau) = 1$. Writing the boundary condition in this fashion is similar to establishing a porous plug at $\xi = 1$, so that the plug catalytically reduces m_A to zero at this point, with the result that the temperature is such that $\tau = 1$ and a finite gradient exists. Establishing a porous plug at $\xi = 1$ essentially divides the problem into the two domains $0 < \xi \leq 1$ and $\xi \geq 1$. Attention is first focused on the $0 < \xi \leq 1$ region.

The case $(d\tau/d\xi) = 0$ must be approached indirectly by interpolation because the profile often degenerates into the $\xi = 0$ and horizontal $\tau = 1$ so that the right-hand side of the equation becomes indeterminate.

Solution of the equation for $K = 0$, the idealized problem without heat loss (flame speed problem), yields a different result for each value of $(d\tau/d\xi)$, so the required value of λ for $(d\tau/d\xi)_1 = 0$ is obtained by extrapolation.

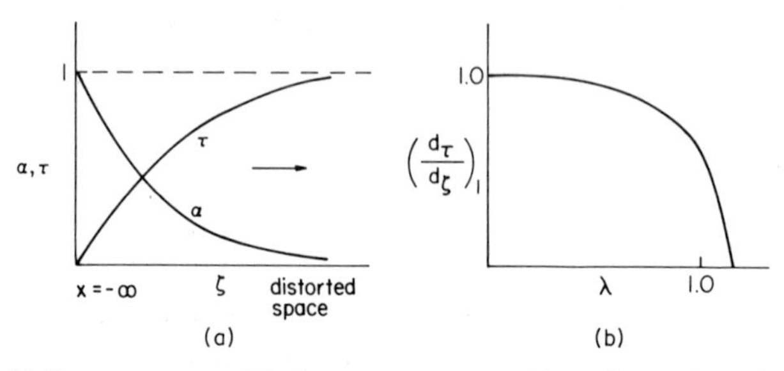

Fig. 18. (a) Temperature and fuel concentration profile in flame plotted as function of distorted space variable. (b) Relationship between temperature gradient at catalyst plug and value of λ.

The trends of the approach are shown in Fig. 18. Extrapolation errors are generally small. Also shown in Fig. 18 are the variations of α and τ with the space variable ξ.

When $K > 0$, the solution differs in that τ falls gradually to 0 as ξ increases beyond the reaction zone; the α and τ are not similar in both equations. The profiles of α and τ take the form shown in Fig. 19.

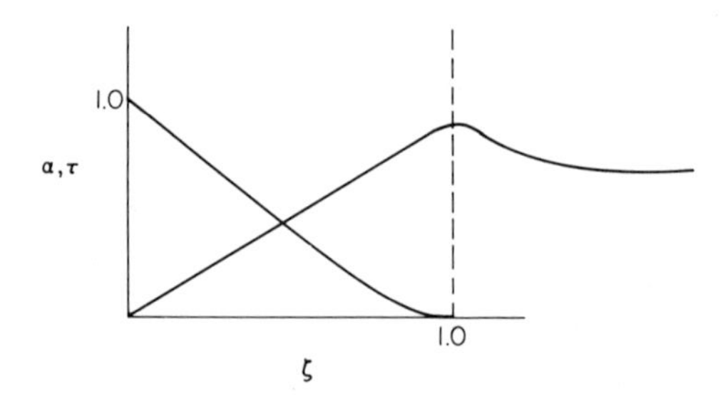

Fig. 19. τ and α profile in a flame with heat losses from burned gases.

The catalyst plug concept is used again since it separates the region in which the reaction rate is dominant from that in which heat losses are dominant. Now attention must be concentrated on $d\alpha/d\xi$, which in general is less than zero. The gradient of τ will then exhibit a discontinuity at $\xi = 1$.

The general solution is complex and must be solved by the method of successive approximations; however, it is possible to examine an approximation for a choice of energy release reactions. This approximate analysis is restricted to reacting gas mixtures which can be characterized by

$$\phi = \alpha\tau^n, \quad \psi = 0 \quad \text{for} \quad \xi < 1, \qquad \psi = \tau^m \quad \text{for} \quad \xi \geq 1 \tag{79}$$

The first of the expressions signifies that the reaction rate is proportional to the A (fuel) concentration. This approximation is close to being true for limit mixtures in which the other reactant B is in large excess. τ^n is a good representation of the rate if n is between 6 and 15. m varies from 1 to 5 according to whether heat transfer is predominantly conduction or radiation.

Upstream of $\xi = 1$, the heat loss term is neglected in comparison with the chemical energy release reaction rate term. This procedure permits the solution to be developed in two parts which must be fitted at $\xi = 1$. Thus for the region $0 < \xi \leq 1$, the equations take the form

$$d^2\tau/d\xi^2 = -\lambda\, \alpha\tau^m/\xi^2, \qquad d^2\alpha/d\xi^2 = +\lambda\, \alpha\tau^m/\xi^2$$

with the boundary conditions

$$\xi = 0, \quad \alpha = 1, \quad \tau = 0; \qquad \xi = 1, \quad \alpha = 0, \quad \tau = \tau_1, \quad d\tau/d\xi = (d\tau/d\xi)_1$$

The equations for α and τ are similar but the boundary conditions are not. Thus α and τ must be related linearly and this relationship can be written as

$$\alpha + \tau = a\xi + b$$

Using the boundary conditions, a and b can be determined in this relationship, to give

$$\alpha = 1 - \tau - (1 - \tau_1)\xi$$

At $\xi = 1$, $\alpha = 0$, so that

$$(d\tau/d\xi)_1 = (\tau_1 - 1)$$

Thus if $\tau_1 = 1$, as before, then $(d\tau/d\xi) = 0$ as before. This last expression shows that with a heat loss, $\tau_1 < 1$ and the slope $(d\tau/d\xi)$ is negative. Also, it follows from the new relationship for α and τ that

$$d\alpha/d\xi = -(1 - \tau_1)$$

At $\xi = 1$, $(d\alpha/d\xi)_1 = 0$.

By multiplying the initial nondimensionalized equation by $(d\tau/d\xi)d\tau_1$, and introducing the linear relationship between α and τ and integrating, one obtains

$$\lambda = \frac{\frac{1}{2}[(d\alpha/d\xi)^2_{\xi=1} - (d\alpha/d\xi)^2_{\xi=0}] + (1 - \tau_1)[(d\alpha/d\xi)_{\xi=1} - (d\alpha/d\xi)_{\xi=0}]}{-\int_0^{\tau_1} \{[1 - \tau - (1 - \tau_1)\xi]/\xi^2\}\tau^n\, d\tau} \qquad (80)$$

The further assumption is made that n is so large that the reaction is confined to the upper temperature levels. This type of assumption has been made in other analyses discussed previously. Under this condition the integrand is finite and ξ may be taken equal to unity without appreciable error. An additional consequence is that at $\xi = 0$,

$$(d\alpha/d\xi)_{\xi=0} \cong -1$$

After $\xi = 1$, $d\alpha/d\xi = 0$; therefore at $\xi = 1$

$$(d\alpha/d\xi)_{\xi=1} = 0$$

Although the integration in the expression for λ (Eq. (80)) is from 0 to τ_1, the only values of τ_1 that have weight are those near 1. Again, this result is due to the assumption that all the reaction is in the highest temperature area. In this high temperature area, ξ is always very close to 1, so that it may be taken as constant and equal to 1. The graphical situation is like that shown in Fig. 18.

Since $\phi = 0$ in the first region,

$$d^2(\tau \quad \text{or} \quad \alpha)/d\xi^2 = 0$$

and therefore

$$d(\tau \quad \text{or} \quad \alpha)/d\xi = \text{const}$$

Thus the lines near $\xi = 0$ in Fig. 18 are straight.

Since $(d\alpha/d\xi) = 0$ at $\xi = 1$ and since τ_1 is close to unity,

$$\lambda = \{2 \textstyle\int_0^{\tau_1} (\tau_1 - \tau)\tau^n \, d\tau\}^{-1}$$

Evaluation of this expression for λ yields

$$\lambda = (n + 1)(n + 2)/2{\tau_1}^{n+2}$$

When τ_1 is known, as for example when heat losses are zero so that $\tau_1 = 1$,

$$\lambda = (n + 1)(n + 2)/2$$

and the flame speed can be evaluated. Recall

$$\lambda = HR_A{}^*k_b/(T_b - T_u)(cG)^2 \qquad \text{and} \qquad cG = c_p \rho_u S_L$$

in previously used nomenclature. Thus

$$S_L = \frac{1}{c_p \rho_u}\left(\frac{k_b H}{(T_b - T_u)} R_A{}^* \frac{1}{\lambda}\right)^{1/2}$$

or since $H/(T_b - T_u) \cong c_p$

$$S_L = \left(\frac{k}{c_p \rho_u} \frac{R_A{}^*}{\rho_u} \frac{1}{\lambda}\right)^{1/2}$$

which is the same type of expression found for the developments of flame propagation made previously.

When heat losses are present, τ_1 must be determined first by considering the region beyond $\xi = 1$. There only the following single equation exists since the catalyst removes α at $\xi = 1$,

$$d^2\tau/d\xi^2 = \lambda(K\tau^m/\xi^2) \tag{81}$$

Equation (81) has a simple solution only if $m = 1$;

$$\tau = \lambda K e^{-\ln \xi}$$

However, if λK is small, τ^m can be treated constant close to $\xi = 1$ and integration of Eq. (81) yields

$$d\tau/d\xi = -\lambda K \tau_1{}^m/\xi$$

By setting $\xi = 1$ and recalling that

$$(d\tau/d\xi)_{\xi=1} = \tau_1 - 1$$

one obtains

$$\lambda K = \tau_1^{-m}(1 - \tau_1)$$

But λ has been found already in terms of τ_1 and n, so

$$K = \frac{2\tau_1{}^{n+2-m}}{(n+1)(n+2)}(1 - \tau_1)$$

Provided $n + 2 > m$, two values of τ_1 between 0 and 1 satisfy this equation for a given K. When K exceeds a critical value K_c, τ_1 becomes imaginary.

By differentiating, critical values of K_c, τ_{1c} and λ_c are obtained. A typical plot of such results is shown in Fig. 20. K_c represents the magnitude

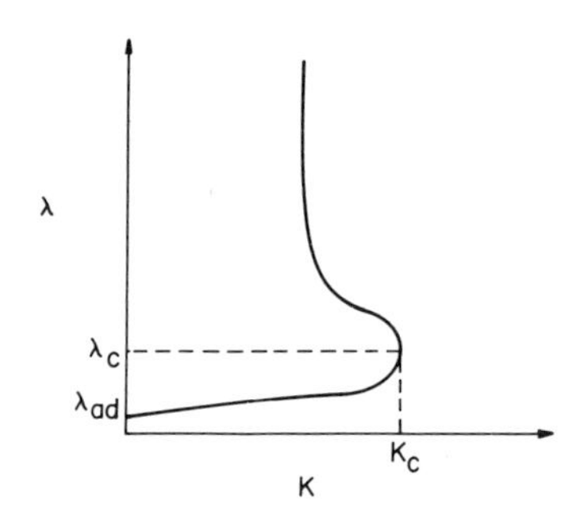

Fig. 20. λ versus K relationship for a particular flame ($n = 11$, $m = 4$).

of the ratio (of the heat loss rate to the chemical reaction energy release rate) which causes a flammability limit.

The flame speed at the limit is determined by the dimensionless parameters and is written as

$$S_{L,c} = (1/c_p \rho_u)(k_b L^*/(T_b - T_u)K_c \lambda_c)^{1/2}$$

To repeat, these flame speeds are at the flammability limits.

The effect of pressure on the flammability limits can be established from the above expression. For heat loss by radiation, $L^* \sim P$. Since $\rho_u \sim P$, then $S_{L,c} \sim P^{-1/2}$. This result is independent of the order of the reaction. But, whether the limit mixture ratio becomes weaker or richer with pressure depends, of course, on the order of the reaction.

It was found in the analysis of flame propagation rates that

$$S_L \sim P^{(n/2)-1}$$

Thus for first-order reactions, S_L varies with pressure in the same manner as $S_{L,c}$ does. So the limit mixture will remain unchanged. The lean limit of a fuel–air mixture could be considered first order, and, indeed, the data given in the previous subsection show the lean limit to be essentially independent of pressure. For second-order reactions, S_L is independent of pressure, but $S_{L,c}$ varies as $P^{-1/2}$. Thus, as the pressure is raised, the limiting flame velocity $S_{L,c}$ drops, and the limits should then broaden as $P^{1/2}$, i.e., the mixture ratio can be made richer at higher pressure so that S_L corresponds to the lower $S_{L,c}$.

The foregoing analysis has been extended to deal with the case in which conduction losses from the flame are greatest, particularly when these losses are from a flame propagating through a tube. This type of flammability limit is called quenching.

In this case, the heat loss is again attributed to the hot gas beyond the reaction zone. The tube is assumed to be at the unburned gas temperature. For a radial heat loss, L can be approximated by

$$L \cong (16\bar{k}/d^2)(T_b - T_u)$$

where $\bar{k}$ is the mean thermal conductivity between T_u and T_b and d is the tube diameter. Thus,

$$L^* = (16\bar{k}^*/d^2)(T_b - T_u)$$

Recall

$$K = L^*/R_A^* H \qquad \text{or} \qquad R_A^* H = L^*/K$$

As before, from the expression of λ_{ad} one may write

$$(G_{ad}c)^2 = \frac{k_b}{T_b - T_u}\frac{L^*}{K\lambda} = \frac{k_b}{(T_b - T_u)\lambda}\frac{16\bar{k}^*}{K}\frac{(T_b - T_u)}{d^2}$$

or

$$\left(\frac{G_{\mathrm{ad}}cd}{k_{\mathrm{u}}}\right)^2 = \frac{16}{K\lambda}\frac{\bar{k}^*k_{\mathrm{b}}}{k_{\mathrm{u}}^2}, \qquad \left(\frac{G_{\mathrm{ad}}cd}{k_{\mathrm{u}}}\right) = \left(\frac{16}{K\lambda}\frac{\bar{k}^*k_{\mathrm{b}}}{k_{\mathrm{u}}^2}\right)^{1/2}$$

For a given mixture K is evaluated, $\lambda = \lambda_{\mathrm{ad}}$, and for a given tube diameter, one can calculate the limiting adiabatic G_{ad} or flame speed.

Since $G_{\mathrm{ad}} = S_{\mathrm{L}}\rho_{\mathrm{u}}$, the above equation shows that the adiabatic flame speed at quenching for a fixed tube diameter is inversely proportional to the pressure, regardless of order of the reaction. Or, the equation shows

$$d_{\mathrm{T}} \sim (k/\rho c_{\mathrm{p}})(1/S_{\mathrm{L}}) \sim 1/P$$

where d_{T} is the symbol for quenching distance. This result is supported by extensive experimental results, as will be shown in the next section. Note, again, that the previous result is independent of the order of the reaction.

2. Quenching Distance

Quenching effects not only alter flammability limits, but also play a role in ignition phenomena which will be discussed in Chapter 7.

The quenching diameter is the parameter given the greatest consideration, and it is generally determined in the following manner. A flame is established on a burner port and the gas mixture flow is suddenly stopped. If the flame propagates down the tube, a smaller tube is substituted until propagation stops. The diameter of the tube which prevents flashback is the quenching distance or diameter.

A flame is quenched in a tube by affecting the two mechanisms which permit flame propagation, i.e., diffusion of species and of heat. Tube walls extract heat; the smaller the tube the greater the surface area to volume ratio within the tube and thus the greater the volumetric heat loss. Similarly, the smaller the tube, the greater the number of collisions of the active radical species with the wall and the greater the number of these species which are destroyed. Since the condition and the material of construction of the wall would affect the rate of destruction of the active species, a specific analytical determination of the quenching distance would be difficult, and the analytical results of the last section must suffice.

Intuition would suggest that there would be an inverse correlation between flame speed and quenching diameter. Since S_{L} varies with ϕ, so should d_{T}; however, the curve for d_{T} would be inverted compared to S_{L}, as shown in Fig. 21.

One would also expect, and it is found experimentally, that increasing

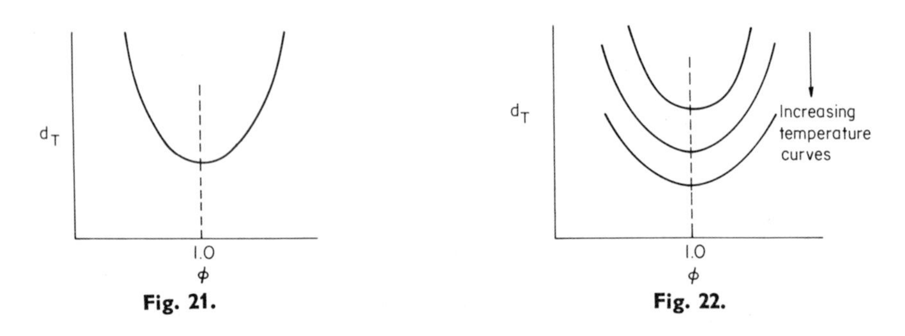

Fig. 21. Variation of quenching diameter with equivalence ratio.

Fig. 22. The effect of temperature on quenching diameter.

the temperature would decrease the quenching distance. This trend could arise due to the fact that heat losses are reduced and species are not as readily deactivated. Sufficient data are not available to develop any specific correlation, but temperature effects would be as depicted in Fig. 22.

It has been concretely established that quenching distance increases as pressure decreases; in fact the correlation is almost exactly

$$d_T \sim 1/P$$

for many compounds. For various compounds, P sometimes has an exponent somewhat less than 1. An exponent close to one in the above relationship can be explained with the following reasoning. The mean free path of gases increases as pressure decreases, thus there are more collisions with the walls and more species are deactivated. Pressure results are generally represented in the form given in Fig. 23, which also establishes that when measuring flammability limits as a function of subatmospheric pressures, the tube diameter must be chosen so that it is greater than the d_T given for the pressure.

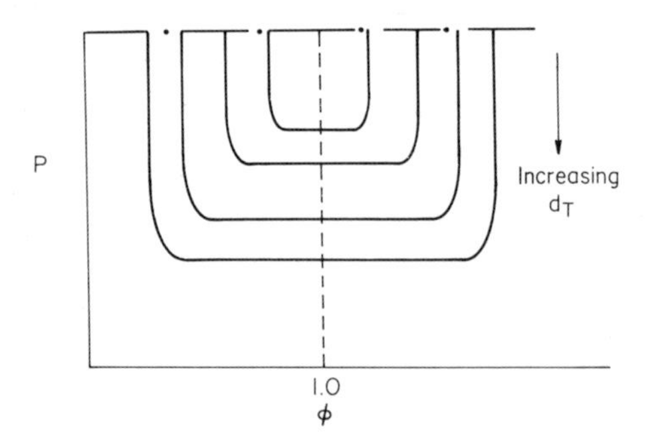

Fig. 23. The effect of pressure on quenching diameter.

The dotted line on Fig. 23 specifies the various flammability limits that would be obtained at a given subatmospheric pressure in tubes of different diameters.

3. Flame Stabilization (Low Velocity)

The mechanism by which the combustion wave maintains a fixed position with respect to a burner rim will be examined next. The Bunsen burner flame can be drawn schematically as shown in Fig. 24. The circled area in Fig. 24 is the anchoring point and the area which is examined in detail. Expanded this area appears as shown in Fig. 25.

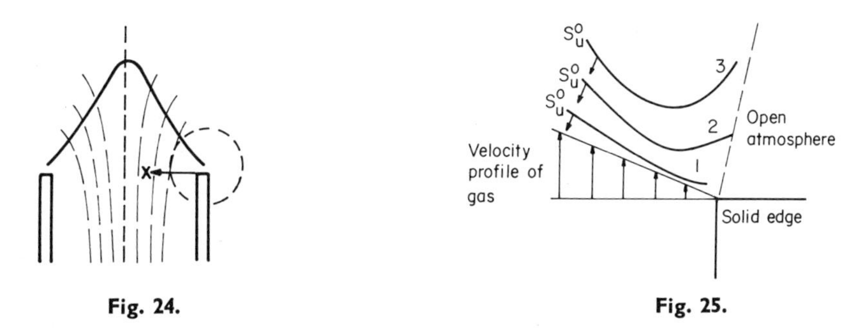

Fig. 24. Streamlines through a Bunsen flame.
Fig. 25. Stabilization position of a Bunsen flame (after Lewis and von Elbe, 1961).

Connected to this flame stabilization idea are the phenomena of flashback, blowoff, and the anchoring of flames on burners. As stated in the last subsection, quenching (for both laminar as well as turbulent flames) also plays a role. If there were no quenching, then flashback could always take place through the boundary layer in the burner tube since the velocity in this boundary layer (really simply the velocity near the wall) becomes lower than the flame speed.

Assuming Poiseuille flow in a laminar jet, the gas velocity is zero at the stream boundary (wall) and increases to a maximum in the center of the stream. The linear dimensions of the region of interest are usually very small. In slow-burning mixtures such as methane and air, they are of the order of 1 mm. Since usually the tube diameter is large in comparison, the gas velocity near the wall is represented by an approximately linear vector profile. It is further assumed that the flow lines of the fuel jet are parallel to the tube axis. Assume that a combustion wave is formed in the stream, and that the fringe of the wave approaches the burner rim closely.

Along the wave profile the burning velocity attains its maximum value S_u^0. Toward the fringe the burning velocity decreases as heat and chain carriers are lost to the rim. If the wave fringe is very close to the rim (position 1 in Fig. 25), the burning velocity in any flow line is smaller than the gas velocity and the wave is driven back by the gas flow. As the distance from the rim increases, the loss of heat and chain carriers decrease and the burning velocity becomes larger. Eventually, a position is reached (position 2 in Fig. 25) in which at some point of the wave profile the burning velocity is equal to the gas velocity. The wave is now in equilibrium with respect to the solid rim. If the wave is displaced to a larger distance (position 3 in Fig. 25), the burning velocity at the indicated point becomes larger than the gas velocity and the wave moves back to the equilibrium position.

a. Flashback

Consider a graph of velocity and distance for a combustion wave inside a tube (Fig. 26). Then allow that for some reason the flame has entered

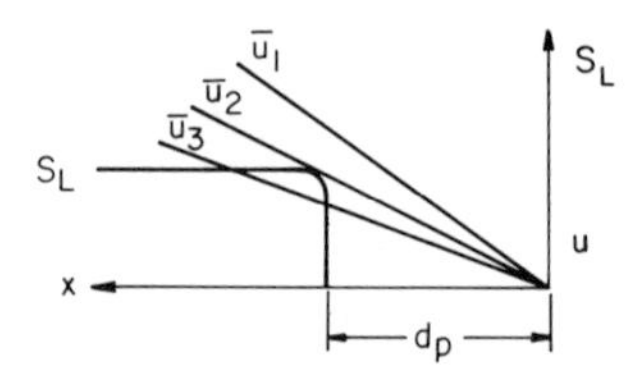

Fig. 26. Burning velocity and gas velocity inside Bunsen tube (after Lewis and von Elbe, 1961).

the tube. The distance between the flame edge and the burner wall is called penetration distance (half the quenching diameter d_T). If u_1 is the mean velocity of the gas and the line labeled u_1 is the graph of the velocity gradient near the rim, then there is no place where the local flame velocity is greater than the local gas velocity; therefore, any flame which finds itself inside the tube will then blow out of the tube. u_2 is then the minimum velocity before flashback. u_3 will flash back.

The gradient for flashback g_F is S_L/d_T. Recall from the analytical development in the section on flammability limits that

$$d_T \approx (\lambda/c_p\rho)(1/S_L)$$

b. Blowoff

Similar reasoning can apply to blowoff, but the arguments are somewhat different and more difficult because nothing similar to a boundary layer exists. But a free boundary does exist.

When the gas flow in the tube is increased, the equilibrium position shifts away from the rim. It is noted that with increasing distance from the rim the gas mixture becomes progressively diluted by interdiffusion with the surrounding atmosphere and the burning velocity in the outermost stream lines decreases correspondingly. This effect is indicated by the increasing retraction of the wave fringe for flame positions 1 to 3 in Fig. 27. But, as the wave moves further from the rim it loses less heat

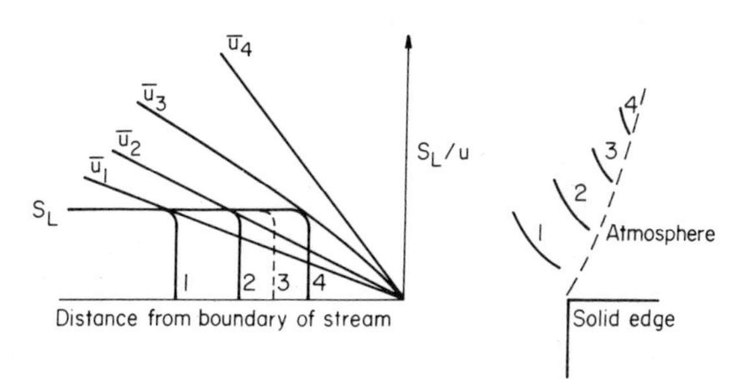

Fig. 27. Burning velocity and gas velocity above a Bunsen tube rim (after Lewis and von Elbe, 1961).

and radicals to the rim so it can extend closer to the hypothetical edge. However, an ultimate equilibrium position of the wave exists beyond which the effect everywhere on the burning velocity of increased distance from the burner rim is overbalanced by the effect of dilution. If the boundary layer velocity gradient is so large that the combustion wave is driven beyond this position, the gas velocity exceeds the burning velocity in every stream line and the combustion wave blows off. These trends are represented diagrammatically in Fig. 27.

The diagram follows the postulated trends in which $S_L{}^0$ is the flame velocity after the gas has been diluted, due to the flame front having moved slightly past u_3. Thus there is blowoff and u_3 is the blowoff velocity.

c. Analysis and Results

Since the topic of concern is the stability of laminar flames fixed to burner tubes, the flow profile of the premixed gases flowing up the tube must be parabolic; i.e., the flow is Poiseuille flow. The gas velocity along any stream is given then by

$$u = n(R^2 - r^2)$$

where R is the tube radius. Since the volumetric flowrate Q cm^3/sec is given by

$$Q = \int_0^R 2\pi r u \, dr$$

then

$$n = 2Q/\pi R$$

The gradient for blowoff or flashback is defined as

$$g_{\mathrm{F,B}} \equiv -\lim_{r \to R} (du/dr)$$

Then

$$g_{\mathrm{F,B}} = 4Q/\pi R^3 = 4\bar{u}_{\mathrm{av}}/R = 8\bar{u}_{\mathrm{av}}/d$$

where d is the diameter of the tube.

Much of the early experimental data on flashback is plotted as a function of the average flashback velocity $u_{\mathrm{av,F}}$ as shown in Fig. 28. It is possible to estimate penetration distance (quenching thickness) from the burner wall in figures such as Fig. 28 by observing the cutoff radius for each mixture.

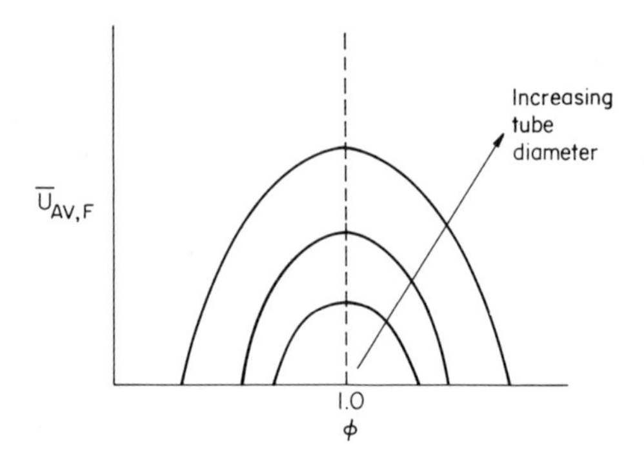

Fig. 28. Critical flow for flashback as a function of equivalence ratio.

The development for the gradients of flashback and blowoff suggests a more appropriate plot of g versus ϕ as shown in Figs. 29 and 30. Examination of these figures reveals that the blowoff curve is much steeper than that of the flashback. For rich mixtures the flashback curves continue to rise instead of decreasing after the stoichiometric value is reached. The reason for this trend is that experiments are performed in air and the diffusion of air into the mixture as the flame lifts off the burning wall increases the local flame speed of the initially fuel rich mixture. Experiments in which the surrounding atmosphere was not air, but nitrogen, verify this explanation and show that the g_{B} would peak at stoichiometric.

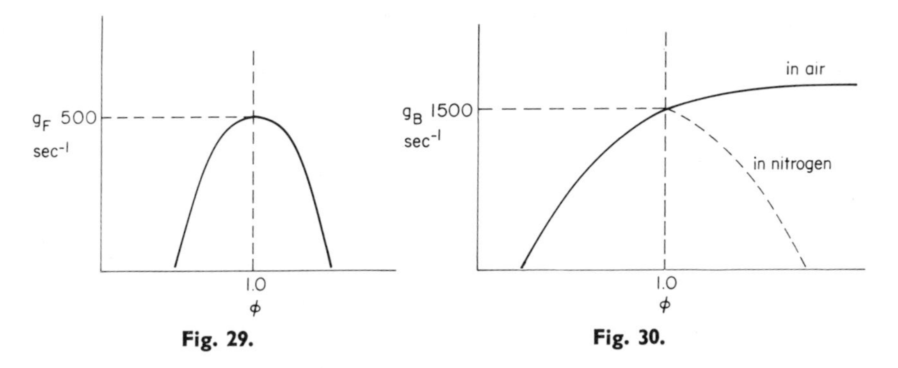

Fig. 29. Typical curve of the gradient of flashback as a function of equivalence ratio. The value at $\phi = 1$ is for natural gas.

Fig. 30. Typical curves of the gradient of blowoff as a function of equivalence ratio. The value at $\phi = 1$ is for natural gas.

4. Stability Limits and Design

The practicality of understanding the stability limits is uniquely obvious when one considers the design of Bunsen tubes and cooking stoves using gaseous fuels.

In the design of a Bunsen burner, it is desirable to have the maximum range of volumetric flow without encountering stability problems. The question arises as to what is the optimum size tube for maximum flexibility. First, the tube diameter must be twice the penetration distance. Second, the average velocity must be at least twice S_L or the Bunsen cone would not form. Experimental evidence shows further that if the average velocity is five times S_L, the fuel penetrates the Bunsen cone tip. If the Reynold's number of the gases in the tube exceeds 2000, the flame becomes turbulent. Of course, there are the limitations of the gradients of flashback and blowoff. If one plots u_{av} versus d for these various limitations, a plot such as that shown in Fig. 31 is obtained. In this figure the dotted region is that in which one can operate and the greatest region of flexibility is about $d = 1$ cm; consequently the tube diameter of Bunsen burners is always about 1 cm.

The burners on cooking stoves are very much like Bunsen tubes. The fuel induces air and the two premix prior to reaching the burner ring with its flame holes. It is possible to idealize this situation as an ejector. For an ejector, the total gas mixture flowrate can be related to the rate of fuel admitted system through conservation of momentum

$$m_m u_m = m_f u_f, \qquad u_m(\rho_m A_m u_m) = (\rho_f A_f u_f) u_f$$

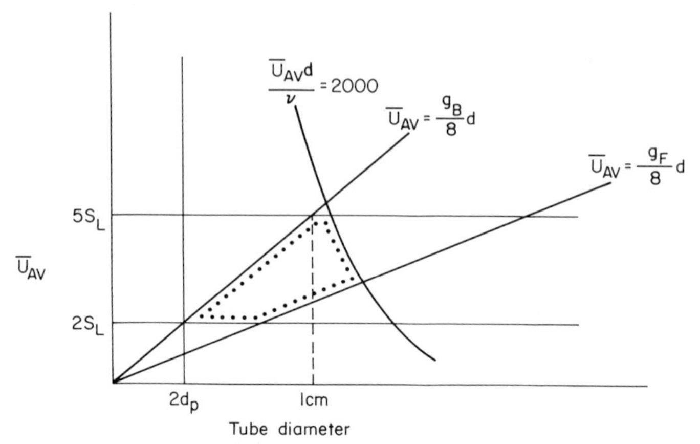

Fig. 31. Stability and operation limits of a Bunsen burner.

where the subscript m represents the conditions for the mixture (A_m is the total area) and the subscript f represents conditions for the fuel. The ejector is depicted in Fig. 32. The momentum expression can be written as

$$\rho_m u_m^2 = \alpha \rho_f u_f^2$$

where α is the area ratio.

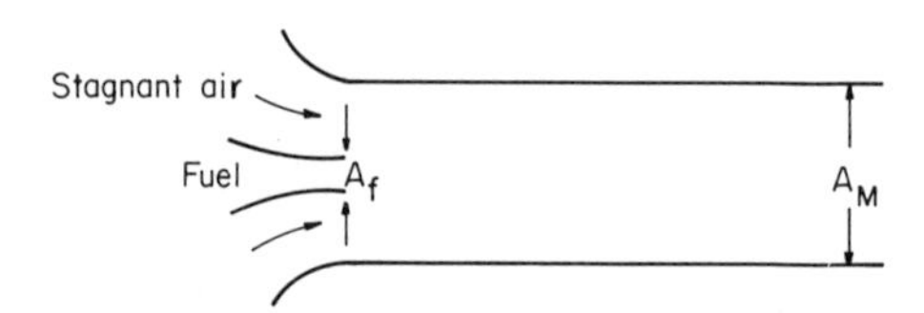

Fig. 32. A fuel-jet ejector system.

If one examines the g_F and g_B on the same graph as shown in Fig. 33, then some interesting observations can be made. The burner port diameter is fixed such that a rich mixture ratio is obtained and at a value represented by the dashed line on Fig. 33. When the mixture ratio is set at this value, the flame can never flashback in the stove and burns without the operator noticing the situation. If the fuel is changed, as the gas industry did when it switched from manufacturer's gas to natural gas or as could possibly occur in the future if the gas companies are forced to switch to a synthetic gas, difficulties could arise.

The volumetric fuel–air ratio in the ejector is given by

$$(F/A) = u_f A_f / u_m A_m$$

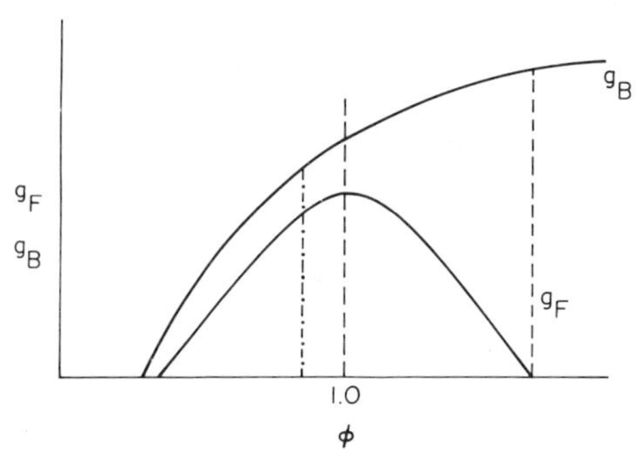

Fig. 33. Flame stability diagram for a gas–air mixture.

It is assumed here that the fuel–air mixture is essentially air. From the momentum equation, this fuel–air mixture ratio becomes

$$(F/A) = (\rho_m/\rho_f)^{1/2}\alpha^{1/2}$$

The stoichiometric molar (volumetric) fuel–air ratio is strictly proportional to the molecular weight of the fuel for two common hydrocarbon fuels, i.e.,

$$(F/A)_{\text{stoich}} \sim 1/mw_f \sim 1/\rho_f$$

The equivalence ratio then is

$$\phi = \frac{(F/A)}{(F/A)_{\text{stoich}}} \sim \frac{\alpha^{1/2}(\rho_m/\rho_f)^{1/2}}{(1/\rho_f)} \sim \alpha^{1/2}\rho_f^{1/2}$$

Figure 33, reveals that for conversion from a heavier fuel to a lighter fuel, the equivalence ratio drops and the dot-dash operating line is obtained. Someone adjusting the same burner with the new lighter fuel would have a very consistent flashback–blowoff problem. Thus when the gas industry switched to natural gas, it was required that every fuel port in every burner on every stove be machined open so that α could become larger to compensate for the decreased ρ_f. Synthetic gases of the future will certainly be heavier than methane (natural gas). They will probably be mostly methane with some heavier components, particularly ethane. Consequently, present burners will not give a stability problem, but will operate more fuel rich and thus be more wasteful of energy. It would be logical to make the fuel ports smaller by caps so that the operating line would be moved next to the flashback rich cutoff line.

C. TURBULENT FLAMES

In the discussions of laminar flame propagation the flames were considered to consist of a smooth, discrete flame zone. Normal and Schlieren photography gave evidence that indeed such is the case. Small flow disturbances may distort the flame and influence the rate of flame propagation but a discrete reaction zone remains. If, however, the unburned gas flow is made turbulent, a diffuse brushy flame results and the rate at which the combustible mixture is consumed increases greatly. The turbulent flame, unlike the laminar flame, is often accompanied by noise and what some people believe are rapid fluctuations of the flame envelope.

For the laminar flame, it is possible to define a flame velocity that, within reasonable limits, is independent of the experimental technique. It would be very desirable to be able to do the same for turbulent flames. Such is not the case, however, and the numerical values of turbulent propagation velocities depend not only on the experimental technique but also on the concept of turbulent flames assumed by the investigator. These points should be kept in mind when considering information presented on turbulent flame velocities and probably represent a concise statement of the state of the field today. The study of turbulent flames is of considerable importance in connection with most practical burner systems. It has been known in a practical way that intense turbulence in the approach flow will effectively increase propagation velocities.

In order to relate the propagation velocity and the combustion intensity to the nature of the approach turbulence by a quantitative theory, the physical structure and basic mechanisms of a turbulent flame must be understood.

The increased rate of burning of a fuel–air mixture in a turbulent flame compared with a laminar flame may be due to any one or a combination of three processes: (1) the turbulent flow may distort the flame so that the surface area is markedly increased, while the normal input of the burning velocity remains the laminar flame velocity (see Fig. 34); (2) turbulence may increase the rate of transport of heat and active species, and thus increase the actual burning velocity normal to the flame surface—and, of course, a reaction zone characterized by special reaction rate functions; and (3) turbulence may rapidly mix the burned and unburned gases in such a way that the flame essentially becomes a homogeneous reaction whose rate depends on the ratio of burned to unburned gases produced in the mixing process.

The first two processes have received the major emphasis, while the third has been considered for some combustor problems. Most theoretical considerations have been based on these concepts and, in general, were considered to have initiated with Damköhler (1940) in Germany and Schelkin (1943) in Russian. Their initial work was continued and extended at other laboratories stimulated by the interest in compact combustors necessary for jet propulsion power plants. In all the early work, both theoretical predictions and experimental observations of turbulent flame speeds were based on the assumption of a continuous wrinkled laminar

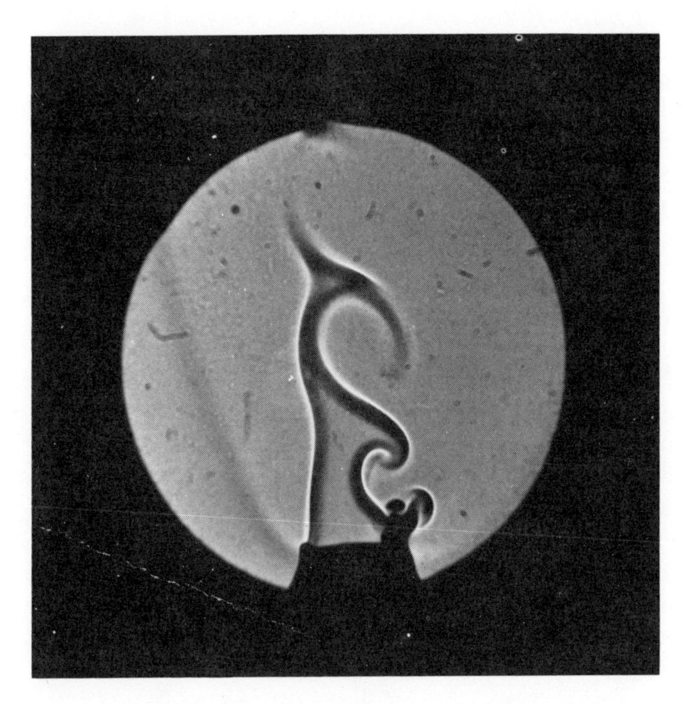

Fig. 34. Flow turbulence induced by a vibrating wire. Spark shadowgraph of 5.6% propane–air flame (after Markstein, 1949).

flame; i.e., it was assumed that the approach flow turbulence distorted but did not disrupt the flame front. This postulate did not result in any clear-cut correlation between experiment and theory, except over a narrow range of flow conditions, i.e., slightly turbulent flow. Further, it was not possible to visualize the physical meaning of a continuous flame front in a fully developed turbulent flame zone.

In 1950, in an effort to remedy the inadequacies of the then existing turbulent flame theories based on the concept of the geometrically distorted laminar figure, workers in the United States and Great Britain introduced the concept of a space heat rate (completely stirred reactors) into the combustion literature. The study of turbulent flame speeds was replaced by the study of volumetric rates of energy release. Theoretical estimates were made on the basis of elementary kinetics of the volumetric rates of energy release which could be expected in combustors in which there was instantaneous mixing of the burned, burning and unburned combustibles. In general, the heat release rates observed in practical devices such as ramjets experimentally were less by a factor of ten than the predicted values.

The calculated values were observed only under conditions corresponding to extreme rates of mixing.

The different investigators were unable to predict either the observed turbulent flame speed or the observed volumetric rates of energy release using a single theory of turbulent combustion. As a result, it has become generally recognized that there must be a spectrum of mechanisms which describe the effect of turbulence on the combustion wave.

1. Work of Damköhler and Schelkin

First, the early theoretical work of Damköhler and Schelkin is examined. Damköhler distinguished between small-scale, high-intensity turbulence in which the eddy size to flame thickness is small and large-scale, small-intensity turbulence in which this ratio is large. He postulated that both situations existed in most flames and stated that large-scale turbulence was of greater importance in combustor application.

He further stated that small-scale turbulence simply increased the transport properties in the wave and investigated these changes as a function of Reynold's number in the following manner. The intensity of turbulence u' may be written as

$$u' = (\overline{u^2})^{1/2}$$

and the scale l as

$$l = u' \int_0^{t'} R_t \, d\tau \qquad \text{where} \quad R_t = \overline{u_0 \, u_\tau}/(u')^2$$

Damköhler pointed out that the eddy diffusivity ε alone may not be sufficient to describe the effects of turbulence on flames since $\varepsilon = lu'$ and both l and u' may have different influences on flame propagation. It is necessary, therefore, to know both the scale and intensity of turbulence.

In the case of large-scale, low-intensity turbulence, the flame will be wrinkled but the laminar transport properties will remain the same; therefore, the laminar flame speed would remain constant. Since for constant S_L, the flame area is proportional to the flow velocity (laminar flames), it would be expected that the increase in area due to turbulence would be proportional to u'. Conceptually, this proportionality may be seen from Fig. 35. Thus for tube flow, the scale l can be considered constant and

$$S_T \sim u' \sim \varepsilon$$

where S_T is the turbulent flame speed. Since $\varepsilon \sim \text{Re}$, then

$$S_T \sim \text{Re}.$$

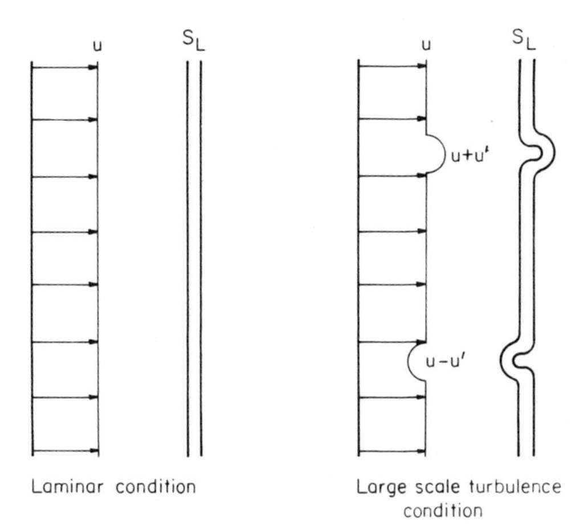

Fig. 35. Distortions of a laminar flame front.

It is not surprising, therefore, that certain experimental results appear to correlate as

$$S_T = A(\mathrm{Re}) + B$$

where A and B are constants.

For small-scale turbulence, a change in the heat transfer and diffusion characteristics should be expected. Damköhler reasoned that the thermal theories give for the laminar flame

$$S_L \sim (\lambda_L/\rho c_p)^{1/2}$$

where λ_L is the molecular thermal conductivity. Since the Prandtl number may be assumed equal to 1, it is possible to write

$$\frac{\lambda}{\rho c_p} \cdot \mathrm{Pr} = \frac{\lambda}{\rho c_p} \cdot \frac{c_p \mu}{\lambda} = \frac{\mu}{\rho} = \nu = \alpha$$

which is explicitly what has been used before, i.e., $\alpha = \nu$. Then,

$$S_L = \nu^{1/2}$$

This line of reasoning was undoubtedly chosen because the turbulent exchange coefficient or eddy viscosity ε has physical significance in fluid mechanics. From similarity, then, Damköhler wrote

$$S_T/S_L = (\varepsilon/\nu)^{1/2}$$

which essentially assumes that there are no other effects from the other variables which arise from the thermal theory of laminar flame propagation.

Schelkin expanded this model by assuming

$$S_L \sim (\lambda_L/\tau)^{1/2}$$

where τ is the reaction time. He assumed that in the turbulent case there was a contribution to the thermal transport from the molecular movements (laminar λ_L) and the eddy movements (λ_T) and wrote

$$S_T \sim [(\lambda_L + \lambda_T)/\tau]^{1/2} \sim \{(\lambda_L/\tau)[1 + (\lambda_T/\lambda_L)]\}^{1/2}$$

or

$$S_T = S_L[1 + (\lambda_T/\lambda_L)]^{1/2}$$

Schelkin also considered large-scale, small-intensity turbulence. He assumed surfaces distorted into cones whose base area was proportional to the square of the average eddy diameter l. The height of the cone is proportional to u' and the time t which an element of the wave is associated with an eddy. This time, then, can be taken equal to (l/S_L). Schelkin then proposed that the ratio of S_T to S_L (average) equals the ratio of the average cone area to the average cone base. From the geometry,

$$A_C = A_B(1 + 4h^2/l^2)$$

where A_C is the area of the cone, A_B the area of the base, and h the cone height. But,

$$h = u't = u'l/S_L$$

therefore,

$$S_T = S_L[1 + (2u'/S_L)^2]^{1/2}$$

For large values of (u'/S_L), the above expression reduces to the expression developed by Damköhler, i.e.,

$$S_T \sim u'$$

An expanded version of the Schelkin concept was presented by Summerfield *et al.* (1955), who wrote

$$S_T/S_L = (\varepsilon/\tau_T)^{1/2}/(\nu/\tau_L)^{1/2}$$

The new feature is that the reaction time is affected by the turbulence. The reaction times can be represented by

$$\tau_L = d_L/S_L, \qquad \tau_T = d_T/S_T$$

where d is the flame thickness (in reality the thickness of the reaction zone). This approach considers the reaction zone to be extended in the turbulent case. Further, conceptually it inherently considers S_L to be a measure of the chemical kinetics alone. With the expressions for τ it follows, then,

$$S_T{}^2/S_L{}^2 = \frac{\varepsilon S_T/d_T}{\nu S_L/d_L}$$

or cross multiplying,

$$S_T\, d_T/\varepsilon = S_L\, d_L/\nu$$

Experimental data on laminar flames give

$$S_L\, d_L/\nu \approx 10$$

thus

$$S_T\, d_T/\varepsilon = S_L\, d_L/\nu \approx 10$$

so that S_T may be determined from the turbulent flame thickness and eddy diffusivity.

2. The Characteristic Time Approach

The assumption that the turbulent flame consists of a continuous laminar flame is explicit in many of the earlier qualitative studies of turbulent combustion. The work of Damköhler and Schelkin had some more recent proponents who, following the same concept of the wrinkled flame, considered the breakup of the flame surface more extensively and thus derived more complicated formulas. Attempts were made to verify this wrinkled flame which, in essence, could be considered a continuous laminar flame by spark Schlieren and electronic probe measurements, but these types of measurements were shown not to give sufficient validity to the concepts.

Many of the early measurements of "turbulent flame speeds" were made by taking photographs of open turbulent flames burning different fuels and subject to different scale and intensities of turbulence. The turbulence characteristics were varied by using burners of different chambers and placing grids in the approach flow. "Turbulent flame speeds" were estimated from the picture of the flame zone and the volumetric gas flow entering the flame zone. Some of the early investigators simply chose the mean area between the inner border observed and the outer. Others took the point of maximum flame intensity, whichever

$$S_T = Q/A$$

S_T is larger than S_L because the turbulent flame is a "stubbier," more compact flame.

Information developed that in many turbulent flames, particularly those in which small-scale, high-intensity turbulence was evident, descrete laminar flamelets do not exist. The failure of this wrinkled flame concept led others to the concept of a distributed reaction zone. Somewhat later, more precise experiments on how turbulence affects flame radiation led to the proposal that there are a series of possible mechanisms which describe the effect of turbulence on the combustion zone.

It has been shown experimentally that changes in equivalence ratio and mixture temperature can cause as much as a tenfold change in the radiant flux emitted by a laminar flame zone. Since mixture ratio changes and preheating may be expected to occur locally in a zone of turbulent flow, it is difficult to understand why the observed reduction due to turbulence is small. In other words, the problem arises of forming a model for a zone of distributed reaction which brings into coincidence the earlier experimental determination that the reaction zone is not merely a distorted laminar flame and the experimental results that the chemistry of the reaction zone does not appear to be greatly affected by turbulence. The radiating radical formation step is thus rapid compared to mixing.

The new postulate, as given by John and Mayer (1957), says that the mechanisms can be interpreted in terms of a characteristic chemical time τ_c and a characteristic aerodynamic time τ_m. The chemical time is defined as

$$\tau_c = d_L/S_L$$

where d_L is the thickness of the laminar flame. τ_c increases with low pressure and decreased chemical activity. The aerodynamic time is defined as

$$\tau_m = \lambda/u'$$

where λ is the microscale of turbulence and, as before, u' is the intensity of turbulence. τ_m increases with decreased flow velocity and increased scale of turbulence.

A nondimensional number can be formed from both these times and has been called the Kovasznay number,

$$\mathrm{Kz} = \tau_c/\tau_m$$

Kovasznay (1956) has written this proportionality in a slightly different form.

Weak turbulence ($\tau_m \gg \tau_c$) merely wrinkles the flame front. In this case τ_m and τ_c are interpreted as being inversely proportional to the velocity gradients characteristic of the flow approaching the flame front and the laminar flame speed, respectively.

Stronger turbulence ($\tau_m \approx \tau_c$) disrupts the laminar flame front. Upon disruption of the flame front, τ_m and τ_c lose their significance as reciprocal velocity gradients. It is postulated that the combustion zone contains a statistical distribution of deflagration centers. In this case, τ_m is interpreted as the mean lifetime of an eddy and τ_c is the time for chemical reaction in these pocket combustibles.

Still stronger turbulence ($\tau \ll \tau_c$) shows its effects by locally diluting and preheating the initial centers of deflagration and on the limit which will result in homogeneous reaction mixtures. Homogeneous reaction in this context is sometimes called the continuously stirred reactor concept and will be discussed in Section D.

3. Experimental Results and Physical and Chemical Effects

Since very few accurate measurements were made of the scale and intensity of the turbulence by those working with turbulent flames, it is most difficult to examine precisely what are the effects of changing the physical and chemical variables in the system. However, some general characteristics of experimental measurements made in turbulent flames are worth reporting.

In laminar flame studies, for a given mixture ratio an increase in volumetric flowrate (i.e., an increase in average tube velocity) leads to an increase in the cone height. In turbulent flame studies, a much weaker change in flame height is observed. This change appears to be a function of the Reynold's number.

The width of a laminar flame is of the order of a millimeter or less. In turbulent flames, the brush width is tens of millimeters. Note from photographs, the turbulent flame appears to be short bushy flames (see Wohl *et al.*, 1957).

If measurements of turbulent flame speeds are made as a function of mixture ratio and the actual value of S_T obtained from Q/A, then the peak value is obtained at the stoichiometric mixture ratio. These observations again point out the strong effect of temperature on flame speed (or in actuality the reaction rate). There is some evidence that the peak S_T shifts to the fuel-rich side as the Reynold's number is increased.

The effect of temperature, or any parameter which affects the temperature, is the same as in laminar flame. Most interestingly, there are no satisfactory data on pressure effects at the present time.

D. STIRRED REACTOR THEORY

In the discussion on premixed turbulent flames, the concept of infinitely fast mixing of the reactants and products was introduced. Generally this concept is referred to as stirred reactor theory and many have applied it not only to turbulent flame phenomena, but also to determine overall reaction kinetic rates and to understand stabilization in high velocity streams (Longwell and Weiss, 1955). Thus the subject is worth reviewing.

Consider a fixed volume V into which fuel and air are injected at a fixed total mass flowrate $\dot{m}$ and temperature T_u. The fuel and air react in the volume and the injection of reactants and outflow of products (also equal to $\dot{m}$) are so oriented that within the volume there is instantaneous mixing of the unburned gases and the reaction products or the burned gases. The reactor volume attains some steady temperature T_R and pressure P. The temperature of the gases leaving the reactor is thus T_R as well. The pressure differential between the reactor and the exit is generally small. The mass leaving the reactor contains the same concentration as within the reactor and thus contains products as well as fuel and air. Within the reactor there exist a certain concentration of fuel (F) and air (A), as well as a fixed unburned mass fraction α. Throughout the reactor volume, T_R, P, (F), (A), and α are constant and fixed; i.e., the reactor is so completely stirred that it is uniform everywhere. Figure 36 is an attempt to depict the stirred reactor concept.

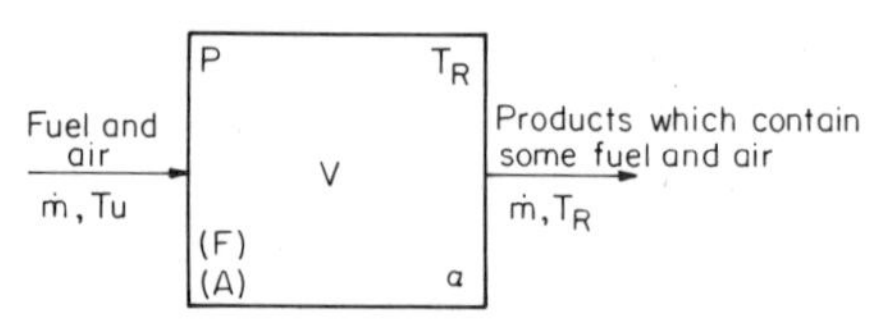

Fig. 36. Variables in a stirred reactor system of fixed volume.

The stirred reactor is to be compared to a plug flow reactor in which premixed fuel–air mixtures flow through the reaction tube. The unburned gases enter at temperature T_u and leave the reactor at the flame temperature T_f, the system is assumed to be adiabatic. Only completely burned products leave the reactor. This reactor is depicted in Fig. 37.

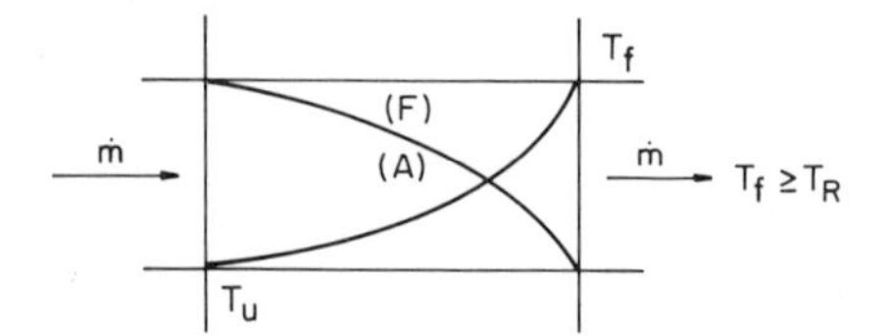

Fig. 37. Variables in a plug flow reactor.

The volume required to convert all the reactants to products for the plug flow reactor is greater than the stirred reactor. The final temperature is, of course, higher than the stirred reactor temperature.

It is relatively straightforward to develop the controlling parameters of a stirred reactor process. α has been defined as the unburned mass fraction. Then it must follow that the mass rate of burning $\dot{R}_B$ is

$$\dot{R}_B = \dot{m}(1 - \alpha)$$

and the rate of heat evolution $\dot{H}$ is

$$\dot{H} = H\dot{m}(1 - \alpha)$$

where H is the heat of reaction per unit mass of reactants for the given fuel–air ratio. If it is assumed that the specific heat of the gases in the stirred reactor can be represented by some average quantity $\bar{c}_p$, then an energy balance may be written as

$$\dot{m}H(1 - \alpha) = \dot{m}\bar{c}_p(T_R - T_u)$$

For the plug flow reactor or any similar adiabatic system, it is also possible to define an average specific heat which takes its explicit definition from

$$\bar{c}_p \equiv H/(T_f - T_u)$$

To a very good approximation the two average specific heats can be assumed equal. Thus it follows that

$$(1 - \alpha) = (T_R - T_u)/(T_f - T_u), \qquad \alpha = (T_f - T_R)/(T_f - T_u)$$

The mass burning rate is determined from the ordinary expression for chemical kinetic rates, i.e.,

$$d(F)/dt = -(F)(A)Z'e^{-E/RT_R} = -(F)^2(A/F)Z'e^{-E/RT_R}$$

where (A/F) represents the air–fuel ratio. The concentration of the fuel can be written in the form

$$(F) = \{(F)/(A) + (F)\}\rho\alpha = \rho\alpha/\{(A/F) + 1\}$$

which permits the rate expression to be written as

$$d(F)/dt = -[(A/F) + 1]^{-2}\rho^2\alpha^2\,(A/F)Z'e^{-E/RT_R}$$

Now the great simplicity in stirred reactor theory is more realizable. (F), (A), and T_R are constant in the reactor and thus the rate of conversion is constant. It is now possible to represent the mass rate of burning in terms of the above chemical kinetic expression:

$$\dot{m}(1 - \alpha) = +V[\{(A) + (F)\}/(F)]\,d(F)/dt$$

or

$$\dot{m}(1-\alpha) = -V[(A/F)+1][(A/F)+1]^{-2}(A/F)\rho^2\alpha^2 Z' e^{-E/RT_R}$$

From the equation of state, by defining

$$B = Z'/[(A/F)+1]$$

and substituting for $(1-\alpha)$ in this last expression, one obtains

$$\left(\frac{\dot{m}}{V}\right) = \left(\frac{A}{F}\right)\left(\frac{\text{PMW}}{RT_R}\right)^2 \left[\frac{(T_f - T_R)^2}{T_f - T_u}\right] \frac{Be^{-E/RT_R}}{[T_R - T_u]}$$

By dividing through by P^2, one observes that

$$(\dot{m}/VP^2) = f(T_R) = f(A/F)$$

This derivation was made as if the overall order of the air-fuel reaction were two. In reality, this order is found to be closer to 1.8. The development could have been carried out for arbitrary overall order n, which would give the result

$$(\dot{m}/VP^n) = f(T_R) = f(A/F)$$

A plot of $(\dot{m}/VP^2)$ versus T_R reveals a multivalued graph which exhibits a maximum as shown in Fig. 38. The part of the curve in Fig. 38 which

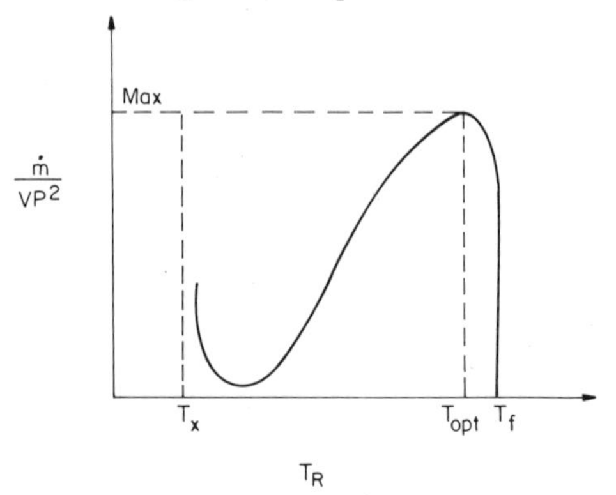

Fig. 38. The stirred reactor parameter $(\dot{m}/Vp^2)$ as a function of reactor temperature T_r.

approaches the value T_x asymptotically cannot exist physically since the mixture could not be ignited at temperatures this low. In fact, the major part of the curve which is to the left of T_{opt} has no physical meaning. At fixed volume and pressure it is not possible for both the mass flowrate and temperature of the reactor to rise. The only stable region is that from T_{opt} to T_f. The reactor parameter must go to zero when $T_R = T_f$ since it is not possible to mix some unburned gases with the product mixture and still obtain the adiabatic flame temperature.

The value of T_R which gives the maximum value of $(\dot{m}/VP^2)$ is obtained by maximizing the last equation. The result is

$$T_{R,\,opt} = T_f/[1 + (2RT_f/E)]$$

For hydrocarbons, the activation energy falls within a range of 30–40,000 cal/mole and the flame temperature in a range of 2000–3000°K. Thus

$$T_{R,\,opt}/T_f \sim 0.75$$

Stirred reactor theory tells us that there is a fixed maximum mass loading rate for a fixed reactor volume and pressure. Any attempts to overload the system will blow out the reaction. Attempts have been made to determine chemical kinetic parameters from stirred reactor measurements. The usefulness of such measurements must be considered limited in nature. First, the analysis is based on the assumption that a hydrocarbon–air system can be represented by a simple overall-order kinetic expression. Recent evidence would indicate that such an assumption is not realistic. Second, the analysis is based on the assumption of instantaneous mixing, which is impossible to achieve experimentally.

In a positive sense, however, it is worthy to note that the analysis does give the maximum overall energy release rate that is possible for a fuel–oxidizer mixture in a fixed volume and at a given pressure.

E. FLAME STABILIZATION IN HIGH VELOCITY STREAMS

The values of laminar flame speeds for hydrocarbon fuels in air are rarely greater than 40 cm/sec. Hydrogen is unique in its flame velocity which approaches 240 cm/sec. If a turbulent flame speed could be attributed to hydrocarbon mixtures, it would at most approach 100 cm/sec. In many practical devices such as ramjet and turbojet combustors, the flow velocities of the fuel–air mixture are of the order 50 m/sec. Further, for such velocities the boundary layers are too thin in comparison to the quenching distance for there to be stabilization by the same means as that which occurs in Bunsen burners. Thus some other means for stabilization is necessary. In practice, the stabilization is accomplished by causing some of the combustion products to recirculate and continually to ignite the fuel mixture. Of course, the continuous ignition could be obtained by inserting small pilot flames. Since pilot flames are an added inconvenience and themselves can blow out, they are generally not used in fast-flowing turbulent streams.

Recirculation of combustion products can be obtained by inserting rods in the stream as pursued in ramjet technology by directing part of the flow normal or against the main stream (aerodynamic stabilization) as is done in gas turbine combustion chambers, or by a step in the wall enclosure.

Photographs of ramjet-type burners which use rods as bluff obstacles show that the regions behind the rods recirculate part of the flow and, indeed, the wake region of the rod acts as a pilot flame. Nicholson and Field (1949) very graphically showed this effect by placing small aluminum particles in the flow (Fig. 39).

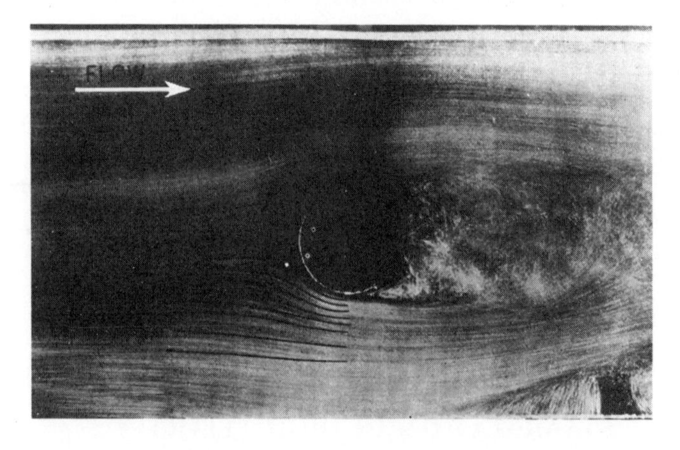

Fig. 39. Flow of 50 ft/sec around $\frac{1}{2}$ in. rod as depicted by aluminum powder technique. Solid lines are flow streamlines of experimenters (Nicholson and Fields, 1949).

The type of obstacle used in stabilization of flames in high speed flows could be gutters, rods, toroids, strips, etc. But in choosing the bluff-body stabilizer the designer must consider pressure drop, cost, ease of manufacture, etc.

Since the combustion chamber should be of minimum length, a single rod, toroid, etc. would be rarely used. In Fig. 40, a schematic of flames spreading from the flame holders is given. Multiple units appreciably shorten the length of the combustion chamber. However, flame holders cause a stagnation pressure loss across the burner and this pressure loss must be added to the large pressure drop due to burning. Thus, there is an optimum between the number of flame holders and pressure drop. It is difficult to use aerodynamic stabilization when large chambers are involved because the

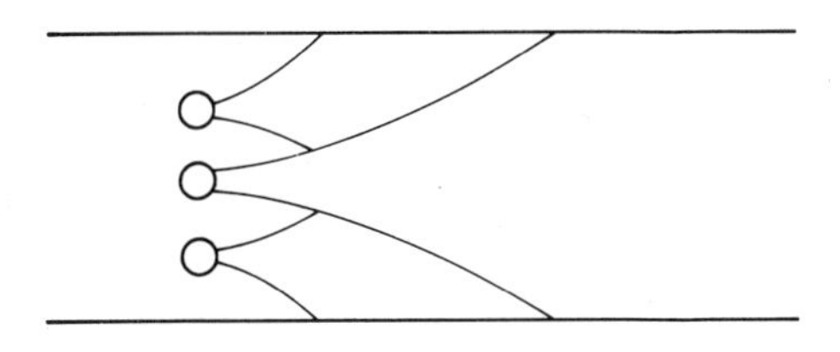

Fig. 40. Flame spreading behind baffles.

flow creating the recirculation would have to penetrate too far across the main stream. Bluff-body stabilization is not used in gas turbine systems because of the combustor shape and length required. In gas turbines a high weight penalty is paid for even the slightest increase in length.

In either case, bluff-body or aerodynamic, the primary concern is that of blow out. In ramjets, the smallest frontal dimension for the highest flow velocity to be used is desirable; in turbojets, it is the smallest volume of the primary recirculation zone that is of concern.

There were many early experimental investigations of bluff-body stabilization. Most of this work used premixed gaseous fuel–air systems and typically plotted the blowoff velocity as a function of the air–fuel ratio for various stabilizer sizes, as shown in Fig. 41. Early attempts to correlate

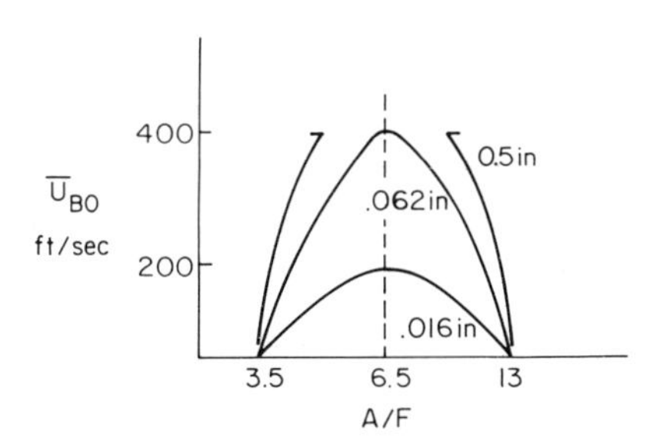

Fig. 41. Blowoff velocities for various rod diameters as a function of air–fuel ratio. Short duct using premixed fuel-air mixtures (after Scurlock, 1948).

the data better appeared to indicate that the dimensional dependency of blowoff velocity was different for different bluff-body shapes. Later it was shown that the Reynold's number range of the experiments was different and that a simple independent dimensional dependency did not exist. Further, the state of turbulence, the temperature of the stabilizer, incoming mixture

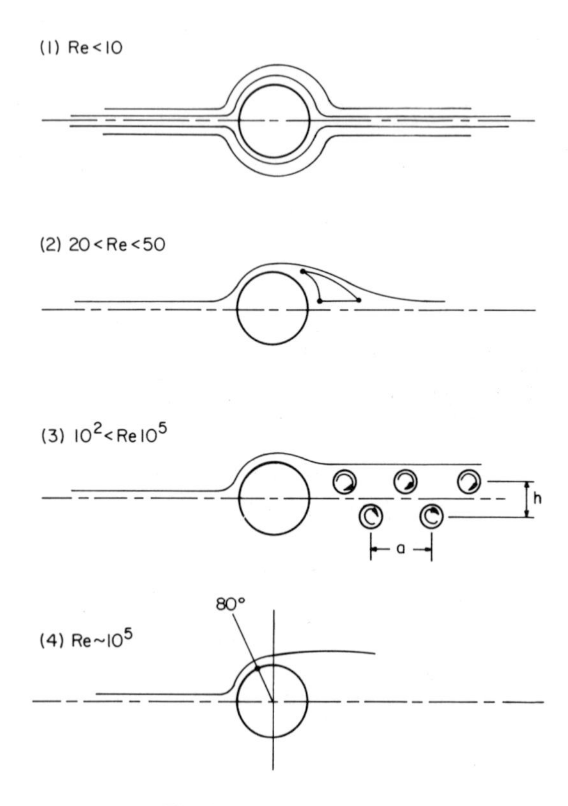

Fig. 42. Flow fields over rods.

temperature, etc. also had secondary effects. All these facts suggest that the fluid mechanics play a significant role in the process.

With the foregoing consideration, it is worth considering the flow field behind a bluff-body (rod) in the region called the wake. Figure 42 depicts the various stages in the development of the wake, as the Reynold's number of the flow is increased. In region (1), there is only slight slowing behind the rod and a very slight region of separation. The heavy dot specifies the stagnation point. In region (2), eddies start to form, and the stagnation points are as indicated. As the Reynold's number increases, the eddy size increases and the downstream stagnation point moves farther away from the rod. In region (3), the eddies become unstable and shed alternately, as shown in the figure. $(h/a) \sim 0.3$. As the velocity V increases, the frequency N of shedding increases, $N \sim 0.3(V/d)$. In region (4), there is a complete turbulent wake behind the body. The stagnation point must pass 90° to about 80° and the boundary layer is turbulent. The turbulent wake behind the body is eventually destroyed downstream by jet mixing.

The discussion of Fig. 42 has been for a two-dimensional rod. For spheres, region (3) does not exist. Also, for combustion behind rods, region (3) does not exist and region (2) extends from $10 < \text{Re} < 10^5$.

With the understanding of the fluid mechanics, two theories were published which correlated the data well. These were due to Spalding (1957) and Zukoski and Marble (1956). A third theory was the adaptation of stirred reactor theory by Longwell *et al.* (1953), who considered the wake zone behind the bluff-body as a stirred-reactor zone.

Zukoski and Marble (1956) considered the wake as an ignition zone, whose length was determined by the characteristic dimension of the stabilizer. At the blowoff condition, the gases flowing past the stabilizer have a contact time equal to the ignition time of the gases; i.e., $\tau_c = \tau_i$, where τ_c is the flow contact time and τ_i is the ignition time. The contact time is given by

$$\tau_c = L/\overline{U}_{bo}$$

where L is the length of the recirculating wake and $\overline{U}_{bo}$ is the velocity at blowoff. Since the length of the wake is proportional to the characteristic dimension of the stabilizer, the diameter d in the case of the rod, then

$$\tau_c \sim d/\overline{U}_{bo}$$

Thus it must follow

$$\overline{U}_{bo}/d \sim 1/\tau_i$$

For second-order reactions, it has been well established that the ignition time is inversely proportional to the pressure. Writing the relation between pressure and time by referencing them to a standard pressure P_0 and time τ_0, one has

$$\tau_0/\tau_i = P/P_0$$

where P is the actual pressure in the system.

The ignition time is a function of the combustion (recirculating) zone temperature which in turn is a function of the air–fuel ratio A/F. Thus

$$\overline{U}_{bo}/dP \sim 1/\tau_0 P_0 = f(T) = f(A/F)$$

Spalding (1955) considers the wake region as one of steady heat transfer with chemical reaction. The energy equation with chemical reaction is considered and nondimensionalized. The solution for the temperature profile along the top of the wake zone is a function of two nondimensional numbers which are functions of one another. There is extinction when these dimensionless groups are not of the same order. Thus the functional extinction condition can be written

$$\overline{U}_{bo}d/\alpha = f(Z'P^{n-1}d^2/\alpha)$$

where d is again the critical dimension, α the thermal diffusivity, Z' the preexponential in the Arrhenius rate constant, and n is the reaction order.

From laminar flame theory, the relationship was obtained that

$$S_L \sim (\alpha RR)^{1/2}$$

or, for purposes here,

$$S_L \sim (\alpha Z' P^{n-1})^{1/2}$$

Since the final expression will be written in terms of the air–fuel ratio which also specifies the temperature, the temperature dependent terms are omitted. Thus a new equality is written

$$S_L{}^2/\alpha \sim Z' P^{n-1}; \qquad Z' P^{n-1} d^2/\alpha = S_L{}^2 d^2/\alpha^2$$

or from above

$$\bar{U}_{\mathrm{bo}} d/\alpha \sim f\{S_L d/\alpha\}$$

Figure 43 shows the overall result obtained by Spalding which correlates data exceptionally well. The different slopes on the figure indicate different Reynold's number regimes of operation. From these slopes the power dependency of d with respect to blowoff velocity can be readily determined.

Longwell *et al.* (1953) essentially stated that the wake was a stirred reactor

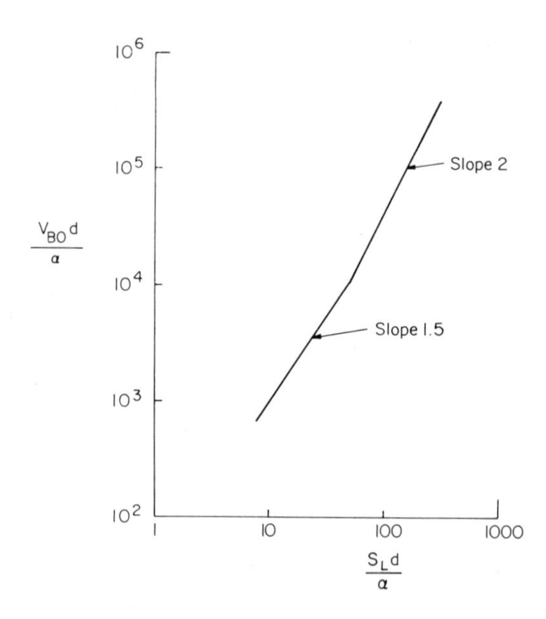

Fig. 43. Correlation of various blowoff velocity data by Spalding (1955).

and that extinction by overloading the reactor was equivalent to blow off of the bluff-body stabilizer. In the previous section for stirred reactors, it was found that

$$\dot{m}/VP^2 = f(A/F)$$

for second-order reactions. $\dot{m}$ is the mass flowrate entering the wake and V is its volume. m and V are then written in the proportional form

$$\dot{m} = \rho A\overline{U} \sim Pd^2\overline{U}_{\mathrm{bo}}, \qquad V \sim d^3$$

where A is an area.

$$Pd^2\overline{U}_{\mathrm{bo}}/d^3P^2 = f(A/F) = \overline{U}_{\mathrm{bo}}/dP$$

This dependency is the same as that found by Zukoski and Marble. Indeed, the Spalding development is of the same form, but it is interesting to note only in the turbulent flow regime where

$$\overline{U}_{\mathrm{bo}}d/\alpha = \mathrm{const}\ (S_{\mathrm{L}}d/\alpha)^2$$

Then

$$\overline{U}_{\mathrm{bo}}/d \sim S_{\mathrm{L}}{}^2/\alpha \sim \alpha RR/\alpha \sim P^{n-1}f(T), \qquad \overline{U}_{\mathrm{bo}}/dP^{n-1} \sim f(T)$$

and thus for a second-order reaction

$$\overline{U}_{\mathrm{bo}}/dP \sim f(T) \sim f(A/F)$$

From these correlations it would be natural to expect that the maximum blowoff velocity as a function of air–fuel ratio would occur at stoichiometric. For the premixed fuel–air systems, the maximum occurs at stoichiometric as shown in Fig. 41. However, in real systems, liquid fuels are injected upstream of the flameholders in order to allow for mixing. Flame stabilization of liquid fuel air systems shows (see May and Maddocks, 1950) that the maximum blowoff velocity is obtained on the fuel-lean side of stoichiometric. This trend is readily explained by the fact that liquid fuel droplets impinge on the stabilizer and enrich the wake. Thus a stoichiometric wake undoubtedly occurs for a lean upstream liquid mixture. That the wake can be modified to alter the blowoff characteristics was shown in very interesting experiments by Fetting *et al.* (1959). The trends of all their results can be explained by the correlations developed in this section.

REFERENCES

Clingman, W. H. Jr., Brokaw, R. S., and Pease, R. D. (1953). *Int. Symp. Combust., 4th*, p. 310. Combustion Inst., Pittsburgh, Pennsylvania.

Coward, H. F., and Jones, O. W. (1951). U.S. Bur. Mines Bull. No. 503.
Damkohler, G. (1947). NACA Tech. Memo. No. 1112.
Fetting, F., Choudhury, A. P. R., and Wilhelm, R. H. (1959). *Int. Symp. Combust. 7th* p. 621. Combustion Inst., Pittsburgh, Pennsylvania.
Fiock, E. S., Marvin, C. S. Jr., Caldwell, F. R., and Roeder, C. H. (1940). NACA Rep. No. 682.
Friedman, R., and Burke, E. (1953). *J. Chem. Phys.* **21,** 710.
Gerstein, M., Levine, O., and Wong, E. L. (1951). *J. Am. Chem. Soc.* **73,** 418.
Hirschfelder, J. O., Curtiss, C. F., and Bud, R. B. (1954). "The Molecular Theory of Gases and Liquids," Chapter 11. Wiley, New York.
John, R. R., and Mayer, E. (1957). Arde Assoc. Tech. Note 4555–5.
Jones, G. W., and Kennedy, R. E. (1945). U.S. Bur. Mines Rep. Invest. No. 3798.
Jones, G. W., Kennedy, R. E., and Spolan, I. (1949). U.S. Bur. Mines Rep. Invest. No. 4557.
Kovasznay, L. S. G. (1956). *Jet Propul.* **26,** 485.
Leason, D. B. (1953). *Int. Symp. Combust. 4th*, p. 369. Combustion Inst., Pittsburgh, Pennsylvania.
Lewis, B., and von Elbe, G. (1961). "Combustion, Flames and Explosions of Gases," 2nd ed., Chapter V. Academic Press, New York.
Longwell, J. P., and Weiss, M. A. (1955). *Ind. Eng. Chem.* **47,** 1634.
Longwell, J. P., Frost, E. E., and Weiss, M. A. (1953). *Ind. Eng. Chem.* **45,** 1629.
Mallard, E., and Le Chatelier, H. L. (1883). *Ann. Mines* **4,** 379.
Markstein, G. H. (1949). *Int. Symp. Combust. 3rd.* p. 162. William and Wilkins, Baltimore, Maryland.
May, W. G., and Maddocks, F. E., Jr. (1950). M.I.T. Meteorol. Rep. 54.
NACA Rep. 1300 (1959). Chapter IV.
Nicholson, H. M., and Fields, J. P. (1949). *Int. Symp. Combust., 3rd* p. 44. Combustion Inst., Pittsburgh, Pennsylvania.
Powling, J. (1949). *Fuel* **28,** 25.
Schelkin, K. I. (1947). NACA Tech. Memo. No. 1110.
Scurlock, A. C. (1948). MIT Fuel Res. Lab. Meteorol. Rep. No. 19.
Semenov, N. N. (1951). NACA Tech. Memo. No. 1282.
Smith, M. L., and Stinson, K. W. (1952). "Fuels and Combustion." McGraw-Hill, New York.
Spalding, D. B. (1955). "Some Fundamentals of Combustion," Chapter 5. Butterworths, London.
Spalding, D. B. (1957). *Proc. Roy. Soc. London A* **240,** 83.
Spalding, D. B., and Botha, J. P. (1954). *Proc. Roy. Soc. London A* **225,** 71.
Summerfield, M., Reiter, S. H. Kebely, V., and Mascolo, R. W. (1955). *Jet Propul.* **25,** 377.
Tanford, C., and Pease, R. N. (1947). *J. Chem. Phys.* **15,** 861.
von Karman, T., and Penner, S. S. (1954). Selected Combustion Problems" (*AGARD Combust. Colloq.*), p. 5. Butterworths, London.
Wohl, K. *et al.* (1957). *Int. Symp. Combust., 6th* p. 333. William and Wilkins, Baltimore, Maryland.
Zebatakis, K. S. (1965). U.S. Bur. Mines Bull. No. 627.
Zukoski, E., and Marble, F. (1955). "Combustion Research and Reviews" (AGARD Colloq.), p. 167. Butterworths, London.

Chapter Five

Detonation

A. INTRODUCTION

Established usage of certain words related to combustion phenomena can be misleading, for what appear to be synonyms are not really so. Before proceeding with the topic of detonation, there will be a slight discourse into the semantics of combustion, with some brief mention of subjects to be covered later.

1. Premixed and Diffusion Flames

The previous chapter covered laminar flame propagation. By inspecting the details of the flow, particularly, high speed or higher Reynold's number flow, it was possible to consider the subject of turbulent flame propagation. These subjects (laminar and turbulent flames) are concerned with gases in the premixed state only. The material presented, generally, is not adaptable to the consideration of the combustion of liquids, solids, or diffusing gases.

Diffusion flames can be either laminar or turbulent and are best described as the combustion state controlled by mixing phenomena, i.e., the diffusion of fuel into oxidizer, or vice versa, until some flammable mixture ratio is reached. According to the state of the turbulence in the individual diffusing species, the situation may be either laminar or turbulent. It will be shown later that there are gaseous diffusion flames, that liquid burning proceeds by a diffusion mechanism, and that the combustion of solids and of some solid propellants falls in this category as well.

2. Explosion, Deflagration, and Detonation

Explosion is a term which corresponds to rapid heat release (or pressure rise). An explosive gas or gas mixture is one which will permit rapid energy release, as compared to most steady, low temperature reactions. Certain gas mixtures (fuel and oxidizer) will not propagate a burning zone or combustion wave. These gas mixtures are said to be outside the flammability limits of the explosive gas.

Depending upon whether the combustion wave is a deflagration or detonation, there are limits of inflammability or detonation.

In general, the combustion wave is considered as a deflagration only, although the detonation wave is another class of the combustion wave. The detonation wave is in all essence a shock wave which is sustained by the energy of the chemical reaction in the highly compressed explosive medium in the wave. Thus, a deflagration is a subsonic wave sustained by a chemical reaction. In the normal sense, it is the general practice to call a combustion wave a flame, so combustion wave, flame and deflagration have been used interchangeably.

It is a very common error to confuse a pure explosion and a detonation. An explosion does not necessarily require the passage of a combustion wave through the exploding medium, whereas an explosive gas mixture must

TABLE 1

Qualitative differences between detonations and deflagration in gases

Ratio	Usual magnitude of ratio	
	Detonation	Deflagration
u_u/c_u [a]	5–10	0.0001–0.03
u_b/u_u	0.4–0.7	4–16
P_b/P_u	13–55	0.98–0.976
T_b/T_u	8–21	4–16
ρ_b/ρ_u	1.4–2.6	0.06–0.25

[a] c_u is the acoustic velocity in the unburned gases. u_u/c_u is the Mach number of the wave.

exist in order to have either a deflagration or a detonation. That is, deflagrations and detonations require rapid energy release, and explosions do not require the presence of a waveform.

The difference between deflagration and detonation may be described qualitatively, but extensively, by Table 1 (from Friedman, 1953).

3. The Onset of Detonation

An explosive medium may support either a deflagration or detonation wave depending upon various conditions, the most obvious being confinement and mixture ratio.

Original studies of the initiation of gaseous detonation have shown no single sequence of events. The primary result of an ordinary thermal initiation appears always to be a flame, which propagates with subsonic speed. Where conditions are such that the flame causes adiabatic compression of the still unreacted material ahead of it, the flame velocity speeds up. In some observations, the speed of the flame seems gradually to rise until it equals that of a detonation wave. Normally, a discontinuous change of velocity is observed from the low flame velocity to the high speed of detonation.

In still other observations, the detonation wave has been observed to originate apparently spontaneously some distance ahead of the flame front. The place of origin has been shown to coincide with the location of a shock wave sent out by the expanding gases of the flame.

The following set of observations do seem to offer a suitable explanation of the thermal initiation of a detonation. This initiation is actually a transition from deflagration to detonation.

A tube containing an explosive gas mixture and having either one or both ends open will permit a combustion wave to propagate when ignited at the open end. This wave attains a steady velocity and does not accelerate to a detonation wave.

If the mixture is ignited at the closed end, a combustion wave is formed, and this wave can accelerate and lead to a detonation. A possible mechanism is as follows. The burned gas products from the initial deflagration have a specific volume of the order of 5–15 times that of the unburned mixture and act as a burned gas piston. Through generation of compression waves, this piston imparts a velocity down the tube to the unburned gases ahead of the flame. Since each preceding compression wave tends to heat the unburned gas mixture somewhat, the sound velocity increases and the succeeding waves catch up with the initial one. Further, the preheating tends to increase the flame speed which then accelerates the unburned gas mixture even further to a point where turbulence is developed in the unburned gas. Then, one obtains a still greater flame velocity, acceleration of the unburned gases, and compression waves. Thus a shock which is strong enough itself to ignite the gas mixture can be formed. The reaction zone behind this shock sends forward a continuous compression wave which keeps the shock front from decaying and a stable detonation is obtained.

The flame regime in a detonation is not different from other flames in that it supplies the sustaining energy. A difference does exist in that the detonation front causes chemical reaction by compression (not by diffusion of heat and species) and thus automatically maintains itself. A further but not essential difference is that this flame burns in highly compressed and preheated gases and burns with extreme rapidity.

A shock wave will cause initiation of detonation directly, whereas other ignition sources (open flames, normal sparks, etc.) will not. After initiation by other sources, the flame that initially propagates is governed by slow diffusion of chain carriers or heat conduction and only under conditions as described before will detonation occur. A plane shock, on the other hand, continuously starts the flame reaction by compression as it travels through fresh layers of explosive gases, and the flame regime behind this shock sustains it immediately as before.

B. THE DETONATION VELOCITY

Although no calculations of a laminar flame speed were made in Chapter 4, it is possible to note from Table 1 that the velocities that would have been obtained would be orders of magnitudes less than that of a detonation wave. Indeed, the laminar flame speeds that have been calculated do show values very small compared to those of detonation. The reason that low velocities are obtained is directly related to the fact that it is initially postulated that the problem is one of low Mach number and that the momentum equation degenerates so that the problem is one of uniform pressure. The low Mach number condition was established from the Hugoniot relation.

The Hugoniot plot also establishes that detonation is a large Mach number phenomena, and one must thus include the momentum equation in obtaining the appropriate solutions.

1. Characterization of the Hugoniot Curve and the Uniqueness of the Chapman–Jouguet Point

Again, the development begins with the consideration of a wave fixed in the laboratory space and the unburned gases flowing into it, as was shown in Fig. 3 of Chapter 4. Since it is most important now to establish

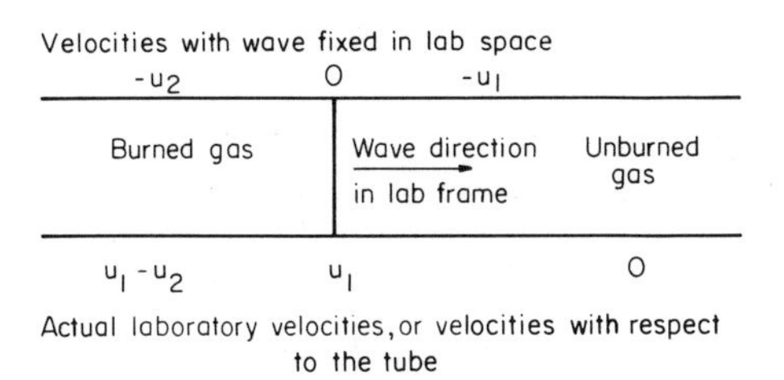

Fig. 1. The velocities in the detonation problem.

a proper understanding of the various velocity symbols to arise, Fig. 3 of Chapter 4 is redrawn (Fig. 1).

The integrated conservation equations and the state equations as before are

$$\rho_1 u_1 = \rho_2 u_2 \tag{1}$$

$$P_1 + \rho_1 u_1{}^2 = P_2 + \rho_2 u_2{}^2 \tag{2}$$

$$c_p T_1 + \tfrac{1}{2} u_1{}^2 + q = c_p T_2 + \tfrac{1}{2} u_2{}^2 \tag{3}$$

$$P_1 = \rho_1 R T_1, \qquad P_2 = \rho_2 R T_2 \qquad \text{(connects known variables)} \tag{4}$$

This type of representation considers that all combustion events are collapsed into a discontinuity (the wave). Thus the unknowns are

$$u_1, \quad u_2, \quad \rho_2, \quad T_2, \quad P_2$$

Since there are four equations and five unknowns, an eigenvalue cannot be obtained. Experimentally it is found that the detonation velocity is uniquely constant for a given mixture. In order to determine all unknowns, it is necessary to know something about the internal structure (rate of reaction), or to obtain another necessary condition.

The method of obtaining a unique solution, or the elimination of many of the possible solutions, will be deferred at present. In order to establish the argument for the nonexistence of various solutions, it is best to pinpoint or define the various velocities that arise in the problem and then to develop certain relationships which will prove convenient.

First, one proceeds to calculate expressions for the velocities u_1 and u_2. From Eq. (1),

$$u_2 = (\rho_1/\rho_2) u_1$$

Substituting in Eq. (2), we have

$$\rho_1 u_1 - (\rho_1{}^2/\rho_2) u_1{}^2 = (P_2 - P_1)$$

Dividing by $\rho_1^{\,2}$, we obtain

$$u_1^{\,2}\left(\frac{1}{\rho_1}-\frac{1}{\rho_2}\right)=\frac{P_2-P_1}{\rho_1^{\,2}},\qquad u_1^{\,2}=\frac{1}{\rho_1^{\,2}}\left[(P_2-P_1)\Big/\left(\frac{1}{\rho_1}-\frac{1}{\rho_2}\right)\right] \tag{5}$$

Since the sound speed c can be written as

$$c_1^{\,2}=\gamma RT_1=\gamma P_1(1/\rho_1)$$

where γ is the ratio of specific heats,

$$\gamma M_1^{\,2}=\left(\frac{P_2}{P_1}-1\right)\Big/\left[1-\frac{(1/\rho_2)}{(1/\rho_1)}\right] \tag{5'}$$

Substituting Eq. (5) into Eq. (1), one obtains

$$u_2^{\,2}=\frac{1}{\rho_2^{\,2}}\left[(P_2-P_1)\Big/\left(\frac{1}{\rho_1}-\frac{1}{\rho_2}\right)\right] \tag{6}$$

and

$$\gamma M_2^{\,2}=\left(1-\frac{P_1}{P_2}\right)\Big/\left[\frac{(1/\rho_1)}{(1/\rho_2)}-1\right] \tag{6'}$$

A relationship which is used throughout these developments is called the Hugoniot equation and is developed as follows. Recall

$$\frac{c_p}{R}=\frac{\gamma}{\gamma-1},\qquad c_p=R\frac{\gamma}{\gamma-1}$$

Substituting in Eq. (3), one obtains

$$R\frac{\gamma}{\gamma-1}T_1+\frac{1}{2}u_1^{\,2}+q=R\frac{\gamma}{\gamma-1}T_2+\frac{1}{2}u_2^{\,2}$$

Implicit in writing the equation in this form is the assumption that c_p and γ are constant throughout. Since $RT=P/\rho$, then

$$\frac{\gamma}{\gamma-1}\left(\frac{P_2}{\rho_2}-\frac{P_1}{\rho_1}\right)-\frac{1}{2}(u_1^{\,2}-u_2^{\,2})=q \tag{7}$$

One then obtains from Eqs. (5) and (6)

$$u_1^{\,2}-u_2^{\,2}=\left(\frac{1}{\rho_1^{\,2}}-\frac{1}{\rho_2^{\,2}}\right)\left[\frac{(P_2-P_1)}{(1/\rho_1)-(1/\rho_2)}\right]=\frac{\rho_2^{\,2}-\rho_1^{\,2}}{\rho_1^{\,2}\rho_2^{\,2}}\left[\frac{(P_2-P_1)}{(1/\rho_1)-(1/\rho_2)}\right]$$

$$=\frac{1}{\rho_1^{\,2}}-\frac{1}{\rho_2^{\,2}}\left[\frac{(P_2-P_1)}{(1/\rho_1)-(1/\rho_2)}\right]=\left(\frac{1}{\rho_1}+\frac{1}{\rho_2}\right)(P_2-P_1) \tag{8}$$

Substituting Eq. (8) into Eq. (7), one obtains the Hugoniot equation

$$\frac{\gamma}{\gamma - 1}\left(\frac{P_2}{\rho_2} - \frac{P_1}{\rho_1}\right) - \frac{1}{2}(P_2 - P_1)\left(\frac{1}{\rho_1} + \frac{1}{\rho_2}\right) = q \tag{9}$$

This relationship, of course, will hold for a shock wave when q is placed equal to zero. The Hugoniot is also written in terms of the enthalpy and internal energy changes. The expression with internal energies is particularly useful in the actual solution for the detonation velocity u_1. If a total enthalpy (sensible plus chemical) in unit mass terms is defined such that

$$h \equiv c_p T + h^\circ$$

where h° is the heat of formation in the standard state and per unit mass, then a simplification of the Hugoniot evolves. Since by this definition

$$q = h_1^\circ - h_2^\circ$$

Eq. (3) becomes

$$\tfrac{1}{2}u_1^2 + c_p T_1 + h_1^\circ = c_p T_2 + h_2^0 + \tfrac{1}{2}u_2^2 \qquad \text{or} \qquad \tfrac{1}{2}(u_1^2 - u_2^2) = h_2 - h_1$$

Or further from Eq. (8), the Hugoniot can also be written as

$$\tfrac{1}{2}(P_2 - P_1)\{(1/\rho_1) + (1/\rho_2)\} = h_2 - h_1 \tag{10}$$

To develop the Hugoniot in terms of the internal energy, one proceeds by first writing

$$h \equiv e + RT = e + P/\rho$$

where e is the total internal energy (sensible plus chemical) per unit mass. Substituting for h in Eq. (10)

$$\frac{1}{2}\left[\left(\frac{P_2}{\rho_1} + \frac{P_2}{\rho_2}\right) - \left(\frac{P_1}{\rho_1} + \frac{P_1}{\rho_2}\right)\right] = e_2 + \frac{P_2}{\rho_2} - e_1 - \frac{P_1}{\rho_1}$$

$$\frac{1}{2}\left[\left(\frac{P_2}{\rho_1} - \frac{P_2}{\rho_2} + \frac{P_1}{\rho_1} - \frac{P_1}{\rho_2}\right)\right] = e_2 - e_1$$

Factoring, another form of the Hugoniot is obtained:

$$\tfrac{1}{2}(P_2 + P_1)\{(1/\rho_1) - (1/\rho_2)\} = e_2 - e_1 \tag{11}$$

There is also interest in the velocity of the burned gases with respect to the tube since as the wave proceeds into the medium at rest it is not known whether the burned gases proceed in the direction of the wave (follow) or proceed away from the wave. From Fig. 1 it is apparent that this velocity, which is also called the particle velocity (Δu), is

$$\Delta u = u_1 - u_2$$

and from Eq. (8)

$$\Delta u = [\{(1/\rho_1) - (1/\rho_2)\}(P_2 - P_1)]^{1/2} \tag{12}$$

Before proceeding further, it must be established which values of the velocity of the burned gases are valid. Thus, it is now best to make a plot of the Hugoniot equation for an arbitrary q. The Hugoniot is essentially a plot of all the possible values of $(1/\rho_2, P_2)$ for a given value of $(1/\rho_1, P_1)$ at the given q. This point $(1/\rho_1, P_1)$ called A is also plotted on the graph.

The regions of possible solutions are constructed by drawing the tangents to the curve that go through $[(1/\rho_1), P_1]$. Since the form of the Hugoniot equation obtained is a hyperbola, there are two tangents to the curve through A as shown in Fig. 2. The tangents and horizontal and vertical

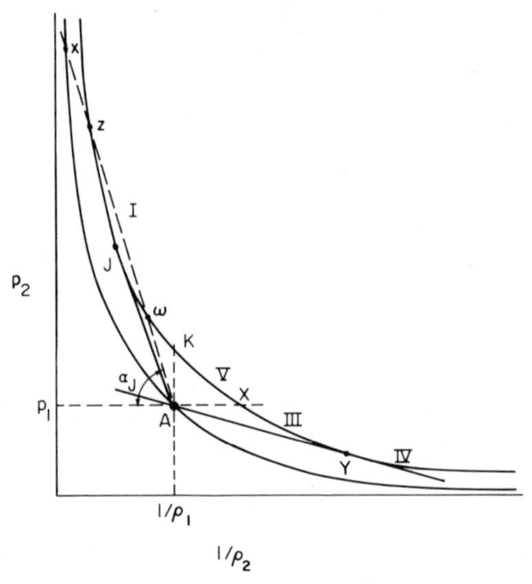

Fig. 2. A Hugoniot plot divided in five regions (I–V).

lines through the initial condition A divide the Hugoniot into five regions, as specified by the Roman numerals. The horizontal and vertical through A are, of course, the lines of constant P and $1/\rho$. A pressure difference for a final condition can be determined very readily from the Hugoniot relation (Eq. (9)) by considering the conditions along the vertical through A; i.e., the condition of constant $(1/\rho)$:

$$\begin{gathered}\frac{\gamma}{\gamma-1}\left(\frac{P_2-P_1}{\rho}\right)-\left(\frac{P_2-P_1}{\rho}\right)=q \\ \left[\left(\frac{\gamma}{\gamma-1}\right)-1\right]\left(\frac{P_2-P_1}{\rho}\right)=q, \qquad (P_2-P_1)=\rho(\gamma-1)q\end{gathered} \tag{13}$$

From Eq. (13), it can be concluded that the pressure differential generated is proportional to the heat release q. If there is no heat release ($q = 0$), $P_2 = P_1$ and the Hugoniot would pass through the initial point A. As inferred before, the shock Hugoniot must pass through A. For different values of q, one obtains a whole family of Hugoniot curves.

The Hugoniot diagram also defines an angle α_J such that

$$\tan \alpha_J = (P_2 - P_1)\bigg/\left(\frac{1}{\rho_1} - \frac{1}{\rho_2}\right)$$

From Eq. (5) then,

$$u_1 = \frac{1}{\rho_1}(\tan \alpha_J)^{1/2}$$

Any other value of α obtained, say by taking points along the curve from J to K and drawing a line through A, is positive and the velocity u_1 is real and possible. However, from K to X, one does not obtain a real velocity due to negative α_J. Thus, region V does not represent real solutions and can be eliminated. A result in this region would require a compression wave to move in the negative direction—an impossible condition.

Regions III and IV give expansion waves, which are the low velocity waves already classified as deflagrations. That these waves are subsonic can be established from the relative order of magnitude of the numerator and denominator of Eq. (6′) and as has already been done in Chapter 4.

Regions I and II give compression waves, high velocities, and are the regions of detonation, as established in Chapter 4.

One can verify that region II gives compression waves and region III expansion waves, by examining the ratio of Δu to u_1 obtained by dividing Eq. (12) by the square root of Eq. (5)

$$\frac{\Delta u}{u_1} = \frac{(1/\rho_1) - (1/\rho_2)}{(1/\rho_1)} = 1 - \frac{(1/\rho_2)}{(1/\rho_1)} \tag{14}$$

In regions I and II, the detonation branch of the Hugoniot curve, $1/\rho_2 < 1/\rho_1$ and the right-hand side of Eq. (14) is positive. Thus in detonations, the hot gases follow the wave. In regions III and IV, the deflagration branch of the Hugoniot, $1/\rho_2 > 1/\rho_1$, and the right-hand side of Eq. (14) is negative. Thus in deflagrations the hot gases move away from the wave.

To this point in the development, the deflagration and detonation branches of the Hugoniot have been characterized and region V has been eliminated. There are some specific characteristics of the tangency point J that were initially postulated by Chapman in 1889. Chapman established that the slope of adiabat is exactly the slope through J, i.e.,

$$\left[\frac{(P_2 - P_1)}{(1/\rho_1) - (1/\rho_2)}\right]_J = -\left[\left(\frac{\partial P_2}{\partial(1/\rho_2)}\right)_s\right]_J \tag{15}$$

The proof of Eq. (15) is a very interesting one and is verified in the following development. From thermodynamics one can write for every point along the Hugoniot

$$T_2\, ds_2 = de_2 + P_2\, d(1/\rho_2) \tag{16}$$

where s is the entropy per unit mass. Differentiating Eq. (11), the Hugoniot in terms of e is

$$de_2 = -\tfrac{1}{2}(P_1 + P_2)\, d(1/\rho_2) + \tfrac{1}{2}[(1/\rho_1) - (1/\rho_2)]\, dP_2$$

since the initial conditions e_1, P_1, and $(1/\rho_1)$ are constant. Substituting this result in Eq. (16),

$$\begin{aligned} T_2\, ds_2 &= -\tfrac{1}{2}(P_1 + P_2)\, d(1/\rho_2) + \tfrac{1}{2}\{(1/\rho_1) - (1/\rho_2)\}\, dP_2 + P_2\, d(1/\rho_2) \\ &= -\tfrac{1}{2}(P_1 - P_2)\, d(1/\rho_2) + \tfrac{1}{2}\{(1/\rho_1) - (1/\rho_2)\}\, dP_2 \end{aligned} \tag{17}$$

It follows from Eq. (17) that along the Hugoniot,

$$T_2\left[\frac{ds_2}{d(1/\rho_2)}\right]_{\mathrm{H}} = \frac{1}{2}\left(\frac{1}{\rho_1} - \frac{1}{\rho_2}\right)\left\{-\frac{(P_1 - P_2)}{(1/\rho_1) - (1/\rho_2)} + \left[\frac{dP_2}{d(1/\rho_2)}\right]_{\mathrm{H}}\right\} \tag{18}$$

The subscript H is used to emphasize that the derivatives are along the Hugoniot curve. Now somewhere along the Hugoniot curve, the adiabatic curve passing through the same point has the same slope as the H curve. ds_2 must be zero there and Eq. (18) gives

$$\left[\left(\frac{dP_2}{d(1/\rho_2)}\right)_{\mathrm{H}}\right]_{\mathrm{s}} = \frac{(P_1 - P_2)}{(1/\rho_1) - (1/\rho_2)} \tag{19}$$

The right-hand side of Eq. (18) is the value of the tangent that also goes through point A; therefore, the tangency point along the H curve is J. Since the order of differentiation on the left-hand side of Eq. (19) can be reversed, it is obvious that Eq. (15) has been developed.

Equation (15) is useful in developing another important condition at point J. The velocity of sound in the burned gas can be written as

$$c_2^2 = \left(\frac{\partial P_2}{\partial \rho_2}\right)_{\mathrm{s}} = -\frac{1}{\rho_2^2}\left[\frac{\partial P_2}{\partial(1/\rho_2)}\right]_{\mathrm{s}} \tag{20}$$

Cross-multiplying and comparing with Eq. (15), one obtains

$$\rho_2^2 c_2^2 = -\left[\frac{\partial P_2}{\partial(1/\rho_2)}\right]_{\mathrm{s}} = \left[(P_2 - P_1)\Big/\left(\frac{1}{\rho_1} - \frac{1}{\rho_2}\right)\right]_J$$

or

$$[c_2^2]_J = \frac{1}{\rho_2^2}\left[(P_2 - P_1)\Big/\left(\frac{1}{\rho_1} - \frac{1}{\rho_2}\right)\right]_J = [u_2^2]_J$$

Therefore

$$[u_2]_J = [c_2]_J \qquad \text{or} \qquad [M_2]_J = 1$$

Thus the important result is obtained that at J the velocity of the burned gases (u_2) is equal to the speed of sound in the burned gases. Further, an exact similar analysis would show, as well, that

$$[M_2]_Y = 1$$

Recall that the velocity of the burned gas with respect to the tube (Δu) is written as

$$\Delta u = u_1 - u_2$$

or at J

$$u_1 = \Delta u + u_2 = \Delta u + c_2 \tag{21}$$

Thus at J the velocity of the unburned gases moving into the wave, i.e., the detonation velocity, equals the velocity of sound in the gases behind the detonation wave plus the mass velocity of this gas (the velocity of the burned gases with respect to the tube). It will be shown presently that this solution at J is the only solution which exists along the detonation branch of the Hugoniot.

Although the complete solution of u_1 at J will not be attempted at this point, it can be shown readily that the detonation velocity has a simple expression, now that u_2 and c_2 have been shown to be equal. The conservation of mass equation is rewritten to show that

$$\rho_1 u_1 = \rho_2 u_2 = \rho_2 c_2 \qquad \text{or} \qquad u_1 = \frac{\rho_2}{\rho_1} c_2 = \frac{(1/\rho_1)}{(1/\rho_2)} c_2$$

Then from Eq. (20) for c_2, it follows that

$$u_1 = \frac{(1/\rho_1)}{(1/\rho_2)} (1/\rho_2) \left\{ -\left(\frac{\partial P_2}{\partial (1/\rho_2)} \right)_s \right\}^{1/2} = \left(\frac{1}{\rho_1} \right) \left\{ -\left(\frac{\partial P_2}{\partial (1/\rho_2)} \right) \right\}_s^{1/2} \tag{22}$$

With the condition that $u_2 = c_2$ at J, it is possible to characterize the different branches in the following manner:

Region I	strong detonation since $P_2 > P_J$ (supersonic flow to subsonic)
Region II	weak detonation since $P_2 < P_J$ (supersonic flow to supersonic)
Region III	weak deflagration since $P_2 > P_Y (1/\rho_2 < 1/\rho_Y)$ (subsonic flow to subsonic)
Region IV	strong deflagration since $P_2 < P_Y (1/\rho_2 > 1/\rho_Y)$

At points above J, $P_2 > P_J$, and thus $u_2 < u_{2,J}$. Since the temperature increases somewhat at higher pressures, then $c_2 > c_{2,J}$ ($c = (\gamma RT)^{1/2}$). Thus M_2 above J must be less than 1. Similar arguments show that points between J and K show that $M_2 > M_{2,J}$ and thus supersonic flow behind the wave. At points past Y, $1/\rho_2 > 1/\rho_Y$, or the velocities are greater than $u_{2,Y}$. Also past Y, the sound speed is about equal to the value at Y. Thus past Y, $M_2 > 1$. A similar argument shows that $M_2 < 1$ between X and Y. Thus past Y, the density decreases, and therefore the heat addition prescribes that there be supersonic outflow. But in a constant area duct, it is not possible to have heat addition and proceed past the sonic condition. Thus region IV is not a physically possible region of solutions and is ruled out.

Region III (weak deflagration) encompasses the laminar flame solutions that were treated in Chapter 4.

There is no condition by which one can rule out strong detonation, but Chapman stated that in this region only velocities corresponding to J were valid. Jouguet (1917) gave the following analysis.

If the final values of P and $1/\rho$ correspond to a point on the Hugoniot curve higher than the point J, it can be shown (next section) that the velocity of sound in the burned gases is greater than the velocity of the detonation wave relative to the burned gases. (It can also be shown that the entropy is a minimum at J and that M_J is greater than values above J.) Consequently, if a rarefaction wave due to any reason whatsoever starts behind the wave, it will catch up with the detonation front. The rarefaction will then reduce the pressure and cause the final value of P_2 and $1/\rho_2$ to drop and move down the curve toward J. Thus points above J are not stable. Heat losses, turbulence, friction, etc., can start the rarefaction. At the point J, the velocity of the detonation wave is equal to the velocity of sound in the burned gases plus the mass velocity of these gases, so that the rarefaction will not overtake it. The point and conditions at J are referred to as the Chapman–Jouguet results.

Thus it appears that solutions in region I are possible, but only in the transient state, since external effects quickly break down this state. Some have claimed to have measured strong detonations in the transient state. There also exist standing detonations which are strong.

The argument which is used to exclude points on the Hugoniot below J is based on the structure of the wave. If a solution in region II were possible, then there would be an equation which would give results both in region I and in region II. A broken line in Fig. 2 representing this equation would go through A and some point, say Z, in region I and another region, say W, in region II. Both Z and W must correspond to the same detonation velocity. The same line would cross the shock Hugoniot at point X. As will be discussed in Section C, the structure of the detonation is a shock wave followed by chemical reaction. Thus to detail the structure of the detonation wave on Fig 2, the pressure could rise from A to X, and then be reduced along the broken line to Z as there is chemical energy release. To proceed to the weak detonation solution at W, there would have to be further energy release. However, all the energy is expended

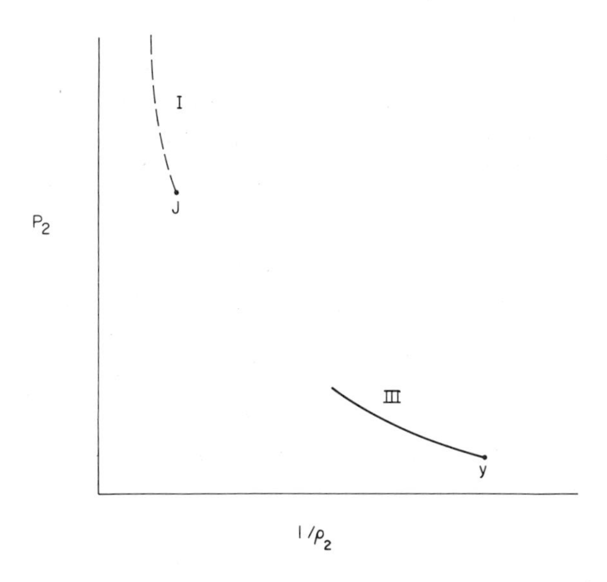

Fig. 3. The only experimentally possible results along the Hugoniot—the point J and region III. The broken line represents transient conditions.

for the initial mixture at point Z. Hence, it is physically impossible to reach the solution given by W as long as the structure requires a shock wave followed by chemical energy release. Therefore, the condition of tangency at J provides the additional condition necessary to specify the detonation velocity uniquely. The physically possible solutions represented by the Hugoniot thus are only those shown in Fig. 3.

2. Determination of the Speed of Sound in the Burned Gases for Conditions above the Chapman–Jouguet Point

a. Behavior of the Entropy along the Hugoniot Curve

Equation (18) was written as

$$T_2\left[\frac{ds_2}{d(1/\rho_2)}\right]_{\mathrm{H}} = \frac{1}{2}\left(\frac{1}{\rho_1} - \frac{1}{\rho_2}\right)\left[\left(\frac{dP_2}{d(1/\rho_2)}\right)_{\mathrm{H}} - \frac{P_1 - P_2}{(1/\rho_1) - (1/\rho_2)}\right] \tag{18}$$

with the further consequence that $(ds_2/d(1/\rho_2)) = 0$ at points Y and J (the latter is the Chapman–Jouguet point).

Differentiating again, and taking into account the fact that

$$(ds_2/d(1/\rho_2)) = 0$$

at point J, one obtains

$$\left[\frac{d^2s}{d(1/\rho_2)^2}\right]_{\text{H at } J \text{ or } Y} = \frac{(1/\rho_1) - (1/\rho_2)}{2T_2}\left[\frac{d^2P_2}{d(1/\rho_2)^2}\right] \tag{23}$$

Now $[d^2P_2/d(1/\rho_2)^2] > 0$ everywhere, since the Hugoniot curve has its concavity directed toward the positive ordinates (see formal proof later).

Therefore at point J, $(1/\rho_1) - (1/\rho_2) > 0$, and hence the entropy is minimum at J. At point Y, $(1/\rho_1) - (1/\rho_2) < 0$, and hence s_2 goes through a maximum.

When $q = 0$, the Hugoniot curve represents an adiabatic shock. Point 1 $(P_1, 1/\rho_1)$ is then on the curve, and Y and J are at 1. Then $(1/\rho_1) - (1/\rho_2) = 0$, and the classical result of the shock theory is found; i.e., *the shock Hugoniot curve osculates the adiabat at the point representing the conditions before the shock.*

Along the detonation branch of the Hugoniot curve the variation of the entropy is as given in Fig. 4. For the adiabatic shock, the entropy variation is as shown in Fig. 5.

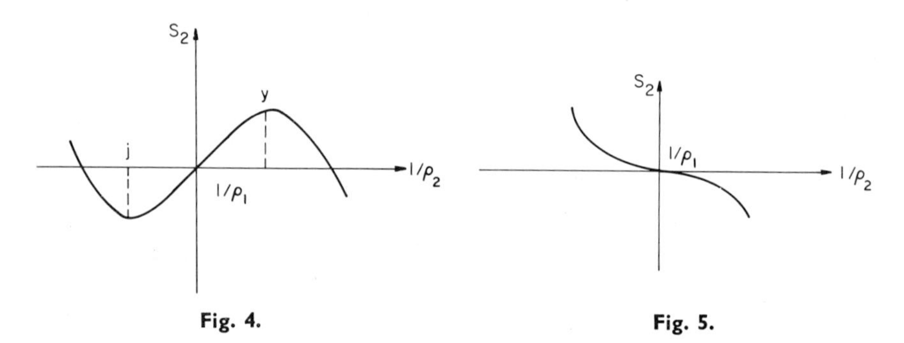

Fig. 4. The variation of entropy along the Hugoniot.
Fig. 5. The entropy variation for the adiabatic shock.

The Concavity of the Hugoniot Curve

Solving for P_2 in the Hugoniot relation, one obtains

$$P_2 = \frac{a + b(1/\rho_2)}{c + d(1/\rho_2)}$$

where

$$a = q + \frac{\gamma + 1}{2(\gamma - 1)} \frac{P_1}{\rho_1}, \qquad b = -\frac{1}{2} P_1, \qquad c = -\frac{1}{2} \rho_1^{-1}, \qquad d = \frac{\gamma + 1}{2(\gamma - 1)} \tag{24}$$

From this equation for the pressure, it is obvious that the Hugoniot curve is a hyperbola. Its asymptotes are the lines

$$\frac{1}{\rho_2} = \left(\frac{\gamma - 1}{\gamma + 1}\right)\left(\frac{1}{\rho_1}\right) > 0, \qquad P_2 = -\frac{\gamma - 1}{\gamma + 1} P_1 < 0$$

The slope is

$$\left(\frac{dP_2}{d(1/\rho_2)}\right)_{\text{H}} = \frac{bc - ad}{(c + d(1/\rho_2))^2}$$

where

$$bc - ad = -\left[\frac{\gamma + 1}{2(\gamma - 1)} q + \frac{P_1}{\rho_1} \frac{\gamma}{(\gamma - 1)^2}\right] < 0 \tag{25}$$

since $q > 0$, $P_1 > 0$, $\rho_1 > 0$. A complete plot of the Hugoniot with its asymptotes would be as shown in Fig. 6. From Fig. 6 it is seen, as could be seen from earlier figures, that the part of the hyperbola representing

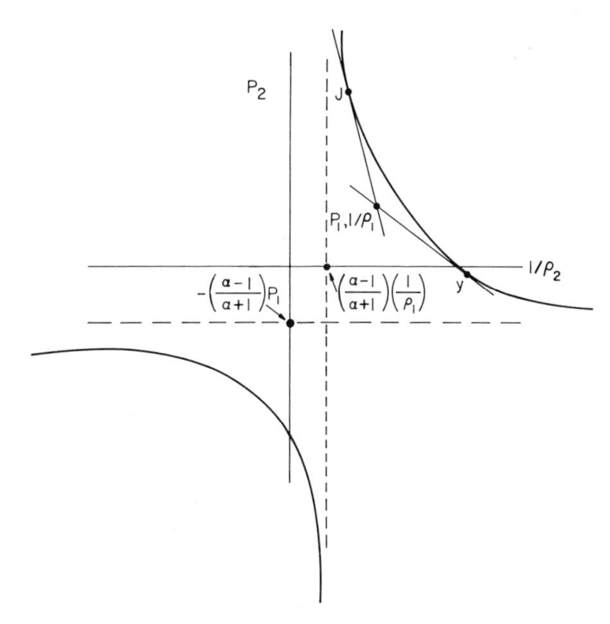

Fig. 6. The asymptotes to the Hugoniot curve.

the strong detonation branch has its concavity directed upwards. It is also possible to determine directly the sign of

$$\left[\frac{d^2P_2}{d(1/\rho_2)^2}\right]_\mathrm{H}$$

by differentiating Eq. (24)

$$\frac{d^2P_2}{d(1/\rho_2)^2} = \frac{2d(ad - bc)}{(c + d(1/\rho_2))^3}$$

Now, $d > 0$, $ad - bc > 0$ (Eq. (25)), and

$$c + d\left(\frac{1}{\rho_2}\right) = \frac{1}{2}\left[\frac{\gamma + 1}{\gamma - 1}\left(\frac{1}{\rho_2}\right) - \left(\frac{1}{\rho_1}\right)\right] > 0$$

The solutions lie on the part of the hyperbola situated on the right-hand side of the asymptote

$$\left(\frac{1}{\rho_2}\right) = \frac{\gamma - 1}{\gamma + 1}\left(\frac{1}{\rho_1}\right)$$

Hence

$$[d^2P_2/d(1/\rho_2)^2] > 0$$

b. The Burned Gas Speed

Since s is a function of P and $1/\rho$,

$$ds = \left(\frac{\partial s}{\partial(1/\rho)}\right)_P d\left(\frac{1}{\rho}\right) + \left(\frac{\partial s}{\partial P}\right)_{1/\rho} dP \tag{26}$$

Since $ds = 0$ for the adiabat Eq. (26) becomes

$$0 = \left(\frac{\partial s}{\partial(1/\rho)}\right)_P + \left(\frac{\partial s}{\partial P}\right)_{1/\rho}\left(\frac{\partial P}{\partial(1/\rho)}\right)_s \tag{27}$$

Differentiating Eq. (26) along the Hugoniot,

$$\left(\frac{ds}{d(1/\rho)}\right)_\mathrm{H} = \left(\frac{\partial s}{\partial(1/\rho)}\right)_P + \left(\frac{\partial s}{\partial P}\right)_{1/\rho}\left(\frac{dP}{d(1/\rho)}\right)_\mathrm{H} \tag{28}$$

Subtracting and transposing Eqs. (27) and (28)

$$\left(\frac{dP}{d(1/\rho)}\right)_\mathrm{H} - \left(\frac{\partial P}{\partial(1/\rho)}\right)_s = \frac{(ds/d(1/\rho))_\mathrm{H}}{(\partial s/\partial P)_{1/\rho}} \tag{29}$$

A thermodynamic expression for the enthalpy is

$$dh = T\,ds + dP/\rho \tag{30}$$

With the conditions of constant c_p and an ideal gas, the expressions

$$dh = c_p\,dT, \qquad T = P/R\rho, \qquad c_p = \{\gamma/(\gamma-1)\}R$$

are developed and substituted in

$$dh = \left(\frac{\partial h}{\partial P}\right) dP + \frac{\partial h}{\partial(1/\rho)}\,d(1/\rho)$$

to obtain

$$dh = \left(\frac{\gamma}{\gamma-1}\right) R\left[\left(\frac{1}{R\rho}\right) dP + \frac{P}{R}\,d\left(\frac{1}{\rho}\right)\right] \tag{31}$$

Combining Eqs. (30) and (31),

$$\begin{aligned} ds &= \frac{1}{T}\left\{\left(\frac{\gamma}{\gamma-1}\right)\frac{1}{\rho}\,dP - \frac{dP}{\rho} + \left(\frac{\gamma}{\gamma-1}\right) P\,d\left(\frac{1}{\rho}\right)\right\} \\ &= \frac{1}{T}\left\{\left(\frac{1}{\gamma-1}\right)\frac{dP}{\rho} + \left(\frac{\gamma}{\gamma-1}\right)\rho RT\,d\left(\frac{1}{\rho}\right)\right\} \\ &= \frac{dP}{(\gamma-1)\rho T} + \left(\frac{\gamma}{\gamma-1}\right) R\rho\left(d\,\frac{1}{\rho}\right) \end{aligned}$$

Therefore

$$(\partial s/\partial P)_{1/\rho} = 1/(\gamma-1)\rho T \tag{32}$$

Then substituting in the values of Eq. (32) into Eq. (29), one obtains

$$\left(\frac{\partial P}{\partial(1/\rho)}\right)_{\mathrm{H}} - \left(\frac{\partial P}{\partial(1/\rho)}\right)_{\mathrm{s}} = (\gamma-1)\rho T\left(\frac{\partial s}{\partial(1/\rho)}\right)_{\mathrm{H}} \tag{33}$$

Equation (18) may be written as

$$\left(\frac{\partial P}{\partial(1/\rho)}\right)_{\mathrm{H}} - \frac{P_1 - P_2}{(1/\rho_1)-(1/\rho_2)} = \frac{2T_2}{(1/\rho_1)-(1/\rho_2)}\left[\frac{\partial s_2}{\partial(1/\rho)}\right]_{\mathrm{H}} \tag{34}$$

Combining Eqs. (33) and (34) gives

$$\begin{aligned} &\left(-\frac{\partial P_2}{\partial(1/\rho_2)}\right)_{\mathrm{s}} - \frac{P_2 - P_1}{(1/\rho_1)-(1/\rho_2)} \\ &\qquad = \left[\frac{\partial s_2}{\partial(1/\rho_2)}\right]_{\mathrm{H}}\left\{-\frac{2T_2}{(1/\rho_1)-(1/\rho_2)} + (\gamma-1)\rho_2 T_2\right\} \end{aligned}$$

or

$$\rho_2{}^2 c_2{}^2 - \rho_2{}^2 u_2{}^2 = \frac{P_2}{R}\left[\gamma - \frac{1+\rho_1/\rho_2}{1-\rho_1/\rho_2}\right]\left[\frac{\partial s_2}{\partial(1/\rho_2)}\right]_{\mathrm{H}} \tag{35}$$

Since the asymptote is given by

$$1/\rho_2 = ((\gamma - 1)/(\gamma + 1))(1/\rho_1)$$

values of $(1/\rho_2)$ on the right-hand side of the asymptote must be

$$1/\rho_2 > ((\gamma - 1)/(\gamma + 1))(1/\rho_1)$$

which leads to

$$\left[\gamma - \frac{(1 + \rho_1/\rho_2)}{(1 - \rho_1/\rho_2)}\right] < 0$$

Since also $(\partial s/\partial(1/\rho_2)) < 0$, then the right-hand side of Eq. (35) is the product of two negative numbers, or a positive number. If the right-hand side of Eq. (35) is positive, then c_2 must be greater than u_2, i.e.,

$$c_2 > u_2$$

3. Calculation of the Detonation Velocity

With the background provided, it is now possible to calculate the detonation velocity for an explosive mixture at given initial conditions. Equation (22)

$$u_1 = \left(\frac{1}{\rho_1}\right)\left[-\frac{dP_2}{d(1/\rho_2)}\right]_s^{1/2}. \tag{22}$$

shows the strong importance of the density of the initial gas mixture, which is reflected more properly in the molecular weight of the products, as will be shown later.

For ideal gases, the adiabatic expansion law is

$$PV^\gamma = \text{const} = P_2(1/\rho_2)^{\gamma_2}$$

Differentiating this expression, one obtains

$$(1/\rho_2)^{\gamma_2}\, dP_2 + P_2(1/\rho_2)^{\gamma_2 - 1}\gamma_2\, d(1/\rho_2) = 0$$

which gives

$$-\left[\frac{dP_2}{d(1/\rho_2)}\right]_S = \frac{P_2}{(1/\rho_2)}\gamma_2 \tag{36}$$

Substituting Eq. (36) into Eq. (22), one obtains

$$u_1 = \frac{(1/\rho_1)}{(1/\rho_2)}[\gamma_2 P_2(1/\rho_2)]^{1/2} = \frac{(1/\rho_1)}{(1/\rho_2)}(\gamma_2 RT_2)^{1/2}$$

If one defines

$$\mu \equiv \frac{(1/\rho_1)}{(1/\rho_2)}$$

then

$$u_1 = \mu(\gamma R T_2)^{1/2} \tag{37}$$

Rearranging Eq. (5), it is possible to write

$$P_2 - P_1 = u_1{}^2 \frac{(1/\rho_1) - (1/\rho_2)}{(1/\rho_1)^2}$$

Substituting for $u_1{}^2$ from above

$$(P_2 - P_1)\frac{(1/\rho_2)}{\gamma_2 P_2} = \left(\frac{1}{\rho_1} - \frac{1}{\rho_2}\right) \tag{38}$$

Now Eq. (11) was

$$e_2 - e_1 = \frac{1}{2}(P_2 + P_1)\left(\frac{1}{\rho_1} - \frac{1}{\rho_2}\right)$$

Substituting Eq. (38) into Eq. (11)

$$e_2 - e_1 = \tfrac{1}{2}(P_2 + P_1)(P_2 - P_1)(1/\rho_2)/\gamma_2 P_2$$

or

$$e_2 - e_1 = \tfrac{1}{2}(P_2{}^2 - P_1{}^2)(1/\rho_2)/\gamma_2 P_2$$

Since $P_2{}^2 \gg P_1{}^2$

$$e_2 - e_1 = \tfrac{1}{2}P_2{}^2(1/\rho_2)/\gamma_2 P_2 = \tfrac{1}{2}P_2(1/\rho_2)/\gamma_2$$

Recall all expressions are in mass units, therefore the gas constant R is not the universal gas constant and, indeed, should now be written R_2 to indicate this condition. Thus

$$e_2 - e_1 = \tfrac{1}{2}P_2(1/\rho_2)/\gamma_2 = \tfrac{1}{2}R_2 T_2/\gamma_2 \tag{39}$$

Recall, as well, that e is the sum of the sensible internal energy plus the internal energy of formation. Equation (39) is the one to be solved in order to obtain T_2 and thus u_1. However, it is more convenient to solve this expression on a molar basis, because the available thermodynamic data and stoichiometric equations are in molar terms.

Equation (39) may be written in terms of the universal gas constant R' as

$$e_2 - e_1 = \tfrac{1}{2}(R'/\mathrm{MW}_2)(T_2/\gamma_2) \tag{40}$$

where MW_2 is the average molecular weight of the products. The gas constant R used throughout this chapter must be the engineering gas constant since all the equations developed were in terms of unit mass, not moles. If one multiplies through Eq. (40) with MW_1, the average molecular weight of the reactants

$$(MW_1/MW_2)e_2(MW_2) - (MW_1)e_1 = \tfrac{1}{2}(R'T_2/\gamma_2)(MW_1/MW_2)$$

or

$$n_2 E_2 - E_1 = \tfrac{1}{2}n_2 R'T_2/\gamma_2 \tag{41}$$

where the E's are the total internal energies and n_2 is (MW_1/MW_2), which is the number of moles of the product per mole of reactant. Usually there are more than one product and one reactant, thus the E's are the molar sums.

Now to solve for T_2, first assume a T_2 and estimate γ_2 and MW_2, which do not vary substantially for burned gas mixtures. For these approximations, it is possible to determine $1/\rho_2$ and P_2.

If Eq. (38) is multiplied by $(P_1 + P_2)$,

$$(P_1 + P_2)\{(1/\rho_1) - (1/\rho_2)\} = (P_2^2 - P_1^2)(1/\rho_2)/\gamma_2 P_2$$

Again $P_2^2 \gg P_1^2$, so that

$$\frac{P_1}{\rho_1} - \frac{P_1}{\rho_2} + \frac{P_2}{\rho_1} - \frac{P_2}{\rho_2} = \frac{P_2(1/\rho_2)}{\gamma_2}$$

$$\left(\frac{P_2}{\rho_1}\right) - \left(\frac{P_1}{\rho_2}\right) = \frac{P_2(1/\rho_2)}{\gamma_2} + \frac{P_2}{\rho_2} - \frac{P_1}{\rho_1}$$

or

$$\frac{P_2\rho_2}{P_2\rho_1} - \frac{P_1\rho_1}{P_1\rho_2} = \frac{R_2T_2}{\gamma_2} + R_2T_2 - R_1T_1$$

$$R_2T_2\left[\frac{(1/\rho_2)}{(1/\rho_2)}\right] - R_1T_1\left[\frac{(1/\rho_2)}{(1/\rho_1)}\right] = \frac{R_2T_2}{\gamma_2} + R_2T_2 - R_1T_1$$

In terms of μ,

$$R_2T_2\mu - R_1T_1(1/\mu) = \{(1/\gamma_2) + 1\}R_2T_2 - R_1T_1$$

which gives

$$\mu^2 - \{(1/\gamma_2) + 1 - (R_1T_1/R_2T_2)\}\mu - (R_1T_1/R_2T_2) = 0 \tag{42}$$

This quadratic equation can be solved for μ; thus for the initial condition $(1/\rho_1)$, $(1/\rho_2)$ is known. It is possible to find P_2, as

$$P_2 = \mu(R_2T_2/R_1T_1)P_1$$

Thus for the assumed T_2, P_2 is known. Then it is possible to determine the equilibrium composition of the burned gas mixture in the same fashion as described in Chapter 1. For this mixture and temperature, both sides of Eq. (39) or (41) are deduced. If the correct T_2 was assumed, both sides of the equation will be equal. If not, reiterate the procedure until T_2 is found. The correct γ_2 and MW_2 will be determined readily. For the correct values, u_1 is determined from Eq. (37) written as

$$u_1 = \mu\left(\frac{\gamma_2 R' T_2}{MW_2}\right)^{1/2}$$

The solution is simpler if the assumption $P_2 \gg P_1$ is made. Then from Eq. (25)

$$\left(\frac{1}{\rho_1} - \frac{1}{\rho_2}\right) = (P_2 - P_1)\frac{(1/\rho_2)}{\gamma_2 P_2} = \frac{P_2(1/\rho_2)}{\gamma_2 P_2}$$

$$\frac{1}{\rho_1} - \frac{1}{\rho_2} = \frac{1}{\gamma_2}\left(\frac{1}{\rho_2}\right), \qquad \left(\frac{1}{\rho_2}\right) = \frac{\gamma_2}{1+\gamma_2}\left(\frac{1}{\rho_1}\right)$$

Since an excellent guess usually can be made of γ_2, one obtains μ immediately and thus P_2.

Gordon and McBride (1971) present a more detailed computational scheme and the associated computational program.

C. THE STRUCTURE OF THE DETONATION WAVE

Zeldovich (1950), Döring (1943) and von Neumann (1942) independently arrived at a theory for the structure of the detonation wave. This theory states that the detonation wave consists of a shock moving at the detonation velocity and leaving heated and compressed gas behind it. The chemical reaction then starts and as it progresses the temperature rises and the density and pressure fall until they reach the Chapman–Jouguet values and the reaction attains equilibrium. A rarefraction wave whose steepness depends on the distance traveled by the wave then sets in. Thus behind the *C–J* shock, energy is generated by thermal reaction.

In the previous section in which the detonation velocity was calculated, the aerodynamic conservation equations were used and it was found that no knowledge of the chemical reaction rate was necessary. The wave was assumed to be a discontinuity. This assumption is satisfactory since these equations placed no restriction on the distance between a shock and the seat of the generating force.

But to look into the structure of the wave one must deal with the kinetics of the chemical reaction. The kinetics and mechanism of reaction give the time and spatial separation of the front and the *C–J* plane.

The distribution of pressure, temperature, and density behind the shock depends upon the fraction of material reacted. If the reaction rate is exponentially accelerating (follows an Arrhenius law), then the fraction reacted changes very little initially and the pressure, density, and temperature profiles are very flat for a distance behind the shock front and then change sharply as the reaction goes to completion at a high rate.

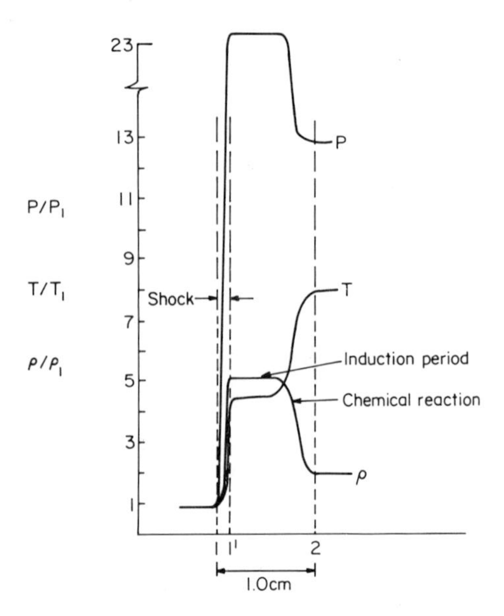

Fig. 7. The variation of physical parameters through a typical detonation wave (see Table 2).

Figure 7 is a graphical representation of the Zeldovich–von Neumann theory and shows the variation of the important physical parameters as a function of spatial distribution. Plane 1 is the shock front, plane 1′ is that immediately after the shock, and plane 2 is the Chapman–Jouguet plane. In the previous section, the conditions for plane 2 were calculated and u_1 obtained. From u_1 and the shock relationships or tables, it is possible to determine the conditions at plane 1′. Typical results for 20% H_2 in air are shown in Table 2.

Thus, as the gas passes from the shock front to the *C–J* state, its pressure drops about a factor of 2, the temperature rises about a factor of 2,

TABLE 2

Calculated values of the physical parameters in a 20% H_2–air detonation

	1	1′	2
M	4.5	0.42	1.0
u	5000 ft/sec	1000	1800
P	1 atm	23	13
T	298°K	1350	2425
ρ/ρ_1	1	5	1.8

and the density drops by a factor of three. It is interesting to follow the model on a Hugoniot plot as shown in Fig. 8.

There are two alternate paths by which a mass element passing through the wave from $\varepsilon = 0$ to $\varepsilon = 1$ may satisfy the conservation equations and at the same time change its pressure and density continuously, not discontinuously, with distance of travel.

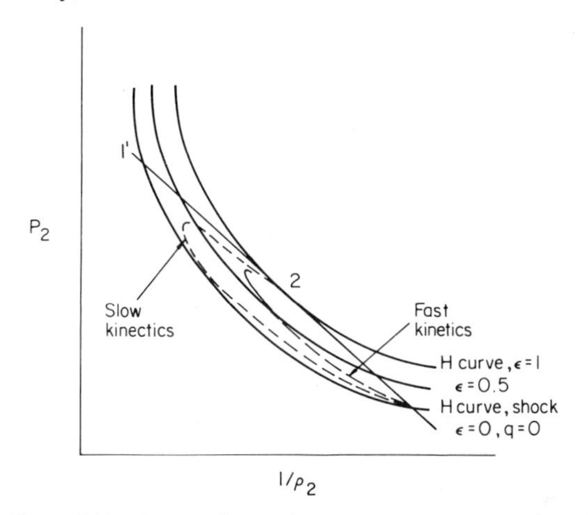

Fig. 8. The effect of kinetics on detonation wave structure as viewed on a Hugoniot curve. ε is the fractional amount of chemical energy converted.

The element may enter the wave in the state corresponding to the initial point and move directly to the *C–J* point. However, this path would demand that reaction occur everywhere along the path. Since there is no compression along this path, there cannot be sufficient temperature to initiate any reaction. Thus there is no energy release to sustain the wave. If on another path a jump is made to the upper point (1′), the pressure and temperature conditions for initiation of reaction are met. In proceeding from 1 to 1′, the pressure does not follow the points along the shock Hugoniot.

The model in which a shock, or at least a steep pressure and temperature rise, creates conditions for reaction and in which the subsequent energy release causes a drop in pressure and density, has been verified by measurements in a detonation tube. Most of these experiments were measurements of the density variation by x-ray absorption.

Hirschfelder *et al.* (1954) solved the complex expressions for the detonation wave where chemical reaction rates are included. They were then able to plot the conditions after region 1 on a Hugoniot. The trends of their results for fast reactions and slow reactions are shown in Fig. 8 also. It is interesting to note that in both cases a pressure between states 1 and 2 is found that is greater than the pressure at state 2.

D. COMPARISON OF DETONATION CALCULATION WITH EXPERIMENTAL RESULTS

In the previous discussion of laminar and turbulent flames, the effects of the physical and chemical parameters on flame speeds were considered and the trends predicted were compared with the experimental trends measured. It is of interest here to recall that it was not possible to calculate these flame speeds explicitly, but as stressed throughout this chapter, it is possible to calculate the detonation velocity accurately. Indeed, the accuracy of the theoretical calculations and the ability to measure the detonation velocity accurately as well has permitted some investigators to calculate thermodynamic properties (such as the bond strength of nitrogen and heat of sublimation of carbon) from experimental measurements of the detonation velocity.

In their book, Lewis and von Elbe (1961) have made numerous comparisons between calculated detonation velocities and experimental values. This book is one of the better sources of such data. Most of the data available for comparison purposes unfortunately were calculated long before the advent of digital computers. Consequently, the theoretical values do not account for all the dissociation that would normally take place. The data presented in Table 3 were abstrated from Lewis and von Elbe and were so chosen to emphasize some important points about the factors which affect the detonation velocity. Although the agreement between the calculated and experimental values in Table 3 can be considered quite good, there is no doubt that the agreement would be much better if dissociation of all possible species was allowed for the final conditions. Modern

TABLE 3

Detonation velocities of stoichiometric hydrogen–oxygen mixture[a]

Mixture	P_2 (atm)	T_2 (°K)	u_1 (m/sec)	
			Calculated	Experimental
$(2H_2 + O_2)$	18.05	3583	2806	2819
$(2H_2 + O_2) + 5O_2$	14.13	2620	1732	1700
$(2H_2 + O_2) + 5N_2$	14.39	2685	1850	1822
$(2H_2 + O_2) + 5H_2$	15.97	2975	3627	3527
$(2H_2 + O_2) + 5He$	16.32	3097	3617	3160
$(2H_2 + O_2) + 5Ar$	16.32	3097	1762	1700

[a] $P_0 = 1$ atm, $T_0 = 291$°K.

computational machinery could certainly perform such calculations. Notice that the calculated values are mostly higher than the experimental values. Further dissociation would have lowered these values quite likely to a greater extent than the difference between the experimental results and the calculations, thus some reactions are frozen.

Variations in the initial temperature and pressure should not affect the detonation velocity for a given initial density. A rise in the initial temperature could only cause a much smaller rise in the final temperature. In laminar flame theory, a small rise in final temperature was important since the temperature was in an exponential term. For detonation theory, recall

$$u_1 = \mu(\gamma_2 R_2 T_2)^{1/2}$$

The subscript 2 has been added in the square root to emphasize that the γ and T are evaluated at the Chapman–Jouguet point and that R is the engineering gas constant.

Also, a change in initial pressure does not affect too severely the final results, since the effect would be in μ which is a specific volume ratio and thus little affected by the pressure. In fact, recall that

$$\mu \cong (\gamma_2 + 1)/\gamma_2$$

γ_2 does not vary significantly.

Examination of Table 3 would lead one to expect that the major factor affecting u_1 would be the initial density. Indeed many have stated that the initial density is one of the most important parameters in determining the detonation velocity. This point is best seen by comparing the results for the mixtures in which the helium and argon inerts are added. The lower density helium mixture gives a much higher detonation velocity than the higher density argon mixture, but identical values of P_2 and T_2 are obtained.

Notice as well that the addition of excess H_2 gives a larger detonation velocity than the stoichiometric mixture. The temperature of the stoichiometric mixture is higher throughout. Again, one could conclude that this variation is a result of the initial density of the mixture. The addition of excess oxygen lowers both the detonation velocity and the temperature. Again, it is possible to argue that excess oxygen increases the initial density.

Whether it is the initial density which is the important parameter should be questioned. The initial density appears in the parameter μ. A change in the initial density by species addition also causes a change in the final density as well, so that, overall, μ does not change appreciably. However, recall that

$$R_2 = R'/\mathrm{MW}_2 \qquad \text{or} \qquad u_1 = \mu(\gamma_2 R' T_2/\mathrm{MW}_2)^{1/2}$$

where R' is the universal gas constant and MW_2 is the average molecular weight of the burned gases. It is really MW_2 that is affected by initial diluents, whether the diluent is an inert or a light weight fuel such as hydrogen. Indeed the ratio of the detonation velocities for the excess helium and the excess argon can be predicted almost exactly, if one takes the square root of the inverse of the ratio of the molecular weights. If it is assumed that there is little dissociation in these two burned gas mixtures, then the reaction products in each case are two moles of water and five moles of helium. In the helium case, the average molecular weight is 9; in the argon case, the average molecular weight is 33.7. The square root of the ratio of the molecular weights is 2.05. The detonation velocity calculated for the argon mixture is 1762. The estimated velocity for helium would be $2.05 \times 1762 = 3560$, which is very close to the calculated result of 3617.

The variation of detonation velocity with mixture ratio is most interesting. Figure 9 shows the variation for acetylene–oxygen mixtures. One notices from Fig. 9 that the maximum detonation velocity is not obtained at the stoichiometric mixture ratio, but the maximum occurs closer to the mixture

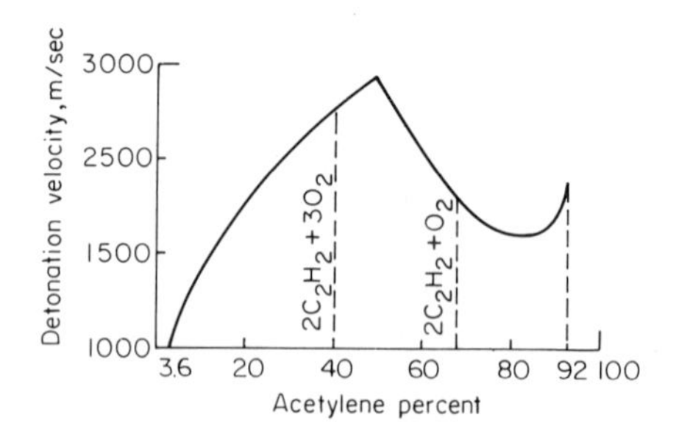

Fig. 9. Detonation velocities of acetylene–oxygen mixtures (after Breton, 1936).

ratio which would correspond to a stoichiometry for the products to be carbon monoxide and water vapor. Indeed the maximum occurs at an even richer mixture ratio. The argument generally given has been that the conversion of CO to CO_2 is too slow to affect the detonation. However, the data of Dryer *et al.* (1971) show that the CO oxidation rates are much faster at high temperature than originally expected. If the previous reasoning were correct, the maximum should occur at the mixture ratio that gives the largest T_2/MW_2. The mixture ratio for acetylene–oxygen which gives a product stoichiometry for CO and H_2 only, is 50% and, indeed, the maximum in Fig. 9 occurs at this value. For propane, the maximum does not occur quite so rich. For the hydrogen–oxygen system, the maximum occurs very near the fuel rich limit.

It is rather interesting to note that the maximum specific impulse of a rocket propellant system occurs when $(T_2/MW_2)^{1/2}$ is maximized, even though the rocket combustion process is not a detonation process.

E. DETONATION LIMITS

Belles (1959) established a very interesting approach to the calculation of detonation limits. An uncertainty in this approach will be discussed, but the line of reasoning developed is worth considering. It is a fine example of coordinating various elements of material presented to this point.

The Belles' prediction of the limits of detonability takes the following course. He deals with the hydrogen–oxygen case. Initially the chemical kinetic conditions for branched-chain explosion in this system are defined in terms of the temperature, pressure, and mixture composition. The standard shock wave equations are used to express, for a given mixture, the temperature and pressure of the shocked gas before reaction is established (condition 1′). The shock Mach number is determined from the detonation velocity. These results are then combined with the explosion condition in terms of M and the mixture composition in order to specify the critical shock strengths for explosion. The mixtures are then examined to determine whether they can support the shock strength necessary for explosion. Some cannot, and they define the limit.

The set of reactions which determine the explosion condition of the hydrogen–oxygen system are

$$OH + H_2 \xrightarrow{k_1} H_2O + H$$

$$H + O_2 \xrightarrow{k_2} OH + O$$

$$O + H_2 \xrightarrow{k_3} OH + H$$

$$H + O_2 + M \xrightarrow{k_4} HO_2 + M$$

The steady state solution shows that

$$d(H_2O)/dt = \text{various terms}/(k_4(M) - 2k_2)$$

consequently the criterion for explosion is

$$k_4(M) = 2k_2 \tag{43}$$

Using rate constant for k_2 and k_4 and expressing the third body concentration (M) in terms of the temperature and pressure by means of the gas law, Belles rewrites Eq. (43) in the form

$$3.11Te^{-8550/T}/f_x P = 1 \tag{44}$$

where f_x is the effective mole fraction of the third bodies in the formation reaction for HO_2. Lewis and von Elbe (1961) give the following empirical relationship for f_x:

$$f_x = f_{H_2} + 0.35f_{O_2} + 0.43f_{N_2} + 0.20f_{Ar} + 1.47f_{CO_2} \tag{45}$$

This expression gives a weighting for the effectiveness of other species as a third body, as compared to H_2 as a third body. Equation (44) is then written as a logarithmic expression

$$(3.710/T) - \log_{10}(T/P) = \log_{10}(3.11/f_x) \tag{46}$$

This equation suggests that if a given hydrogen–oxygen mixture, which could have a characteristic value of f dependent on the mixture composition, is raised to a temperature and pressure that satisfy the equation, then the mixture will be explosive.

For the detonation waves, the following relationships for the temperature and pressure can be written for the condition (1′) behind the shock front. It is these conditions which initiate the deflagration state in the detonation wave:

$$P_{1'}/P_0 = (1/\alpha)[(M^2/\beta) - 1] \tag{47}$$

$$T_{1'}/T_0 = [(M^2/\beta) - 1][\beta M^2 + (1/\gamma)]/\alpha^2\beta M^2 \tag{48}$$

where M is the Mach number, $\alpha = (\gamma + 1)/(\gamma - 1)$, and $\beta = (\gamma - 1)/2\gamma$. Shock strengths in hydrogen mixtures are sufficiently low so that one does not have to be concerned with the real gas effects on the ratio of specific heats γ, and γ can be evaluated at the initial conditions.

From Eq. (46) it is apparent that many combinations of pressure and temperature will satisfy the explosive condition. However, if the condition is specified that the ignition of the deflagration state must come from the shock wave, Belles argues that there is only one Mach number which will satisfy the explosive condition. This Mach number is called the critical

Mach number and is found by substituting Eqs. (47) and (48) into Eq. (46) to give

$$\frac{3.710\alpha^2\beta M^2}{T_0((M^2/\beta)-1)(\beta M^2+(1/\gamma))}-\log_{10}\left[\frac{T_0(\beta M^2+(1/\gamma))}{P_0\,\alpha\beta M^2}\right]=f(T_0,P_0,\gamma,M)$$
$$=\log_{10}(3.11/f_x) \qquad (49)$$

This equation is most readily solved by plotting the left-hand side as a function of M for the initial conditions. The logarithm term on the right-hand side is calculated for the initial mixture and M found from the plot.

The final criterion that establishes the detonation limits is imposed by energy considerations. The shock provides the mechanism whereby the combustion process is continually sustained; however, the energy to drive the shock, i.e., to heat up the unburned gas mixture, comes from the ultimate energy release in the combustion process. But if the enthalpy increase across the shock which corresponds to the critical Mach number is greater than the heat of combustion, an impossible situation arises. No explosive condition can be reached, and the detonation cannot propagate. Thus the criterion for the detonation of a mixture is

$$\Delta h_s \le \Delta h_c$$

where Δh_c is the heat of combustion per unit mass for the mixture and Δh_s is the enthalpy rise across the shock for the critical number (M_c),

$$h_{T_{1'}} - h_{T_0} = \Delta h_s \qquad \text{where} \quad T_{1'} = T_0(1+\tfrac{1}{2}(\gamma-1)M_c^{\,2})$$

The plot of Δh_c and Δh_s for the hydrogen–oxygen case as given by Belles is shown in Fig. 10. Where the curves cross in Fig. 10, $\Delta h_c = \Delta h_s$ and the limits specified. The comparisons with experimental data are very good as is shown in Table 4.

Questions have been raised about this approach to calculate detonation limits, and some believe the general agreement between experiments and the

TABLE 4

Hydrogen detonation limits in oxygen and air

	Lean limit, vol %		Rich limit, vol %	
System	Experimental	Calculated	Experimental	Calculated
H_2-O_2	15	16.3	90	92.3
H_2-Air	18.3	15.8	59.9	59.7

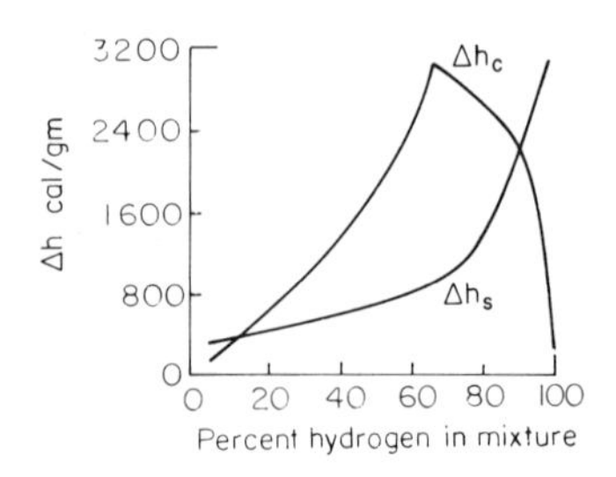

Fig. 10. Heat of combustion per unit mass (Δh_s) as a function of hydrogen concentration for detonation conditions (after Belles, 1959).

theory as shown in Table 4 is fortuitous. One of the criticisms is that a given Mach number specifies a particular temperature and a pressure behind the shock. The expression representing the explosive condition also specifies a particular pressure and temperature. It is unlikely that there would be a direct correspondence of the two conditions from the different shock and explosion relationships. Equation (36) must give a unique result for the initial conditions because of the manner in which it was developed.

Detonation limits have been measured for various fuel–oxidizer mixtures. These values and comparison with the deflagration (flammability) limits are given in Table 5. It is interesting to note that the detonation limits are

TABLE 5

Comparison of deflagration and detonation limits

	Lean		Rich	
	Deflagration	Detonation	Deflagration	Detonation
H_2–O_2	4	15	94	90
H_2–Air	4	18	74	59
CO–O_2	16	38	94	90
NH_3–O_2	15	25	79	75
C_3H_8–O_2	2	3	55	37

always narrower than the deflagration limit. But for H_2 and the hydrocarbons, one should recall that because of the product molecular weight the detonation velocity has its maximum near the rich limit. The deflagration velocity maximum is always very near the stoichiometric value and indeed has its minimum values at the limits. Indeed, the experimental definition of the deflagration limits would require this result.

REFERENCES

Belles, F. E. (1959). *Int. Symp. Combust., 7th* p. 745. Butterworths, London.
Breton, J. (1936). *Ann. Office Nat. Combust. Liquides* **11**, 4871.
Chapman, D. L. (1899). *Phil. Mag.* **47**, 90.
Döring, W. (1943). *Ann. Phys.* **43**, 421.
Dryer, F. L., Naegeli, D. W., and Glassman, I. (1971). *Combust. Flame* **17**, 270.
Friedman, R. (1953). *Am. Rocket Soc. J.* **23**, 349.
Gordon, S., and McBride, B. V. (1971). NASA SP-273.
Hirschfelder, J. O., Curtiss, C. F., and Bird, R. B. (1954). "The Molecular Theory of Gases and Liquids." Wiley, New York.
Jouguet, E. (1917), "Méchanique des Esplosifs." Dorn, Paris.
Lewis, B., and von Elbe, G. (1961). "Combustion, Flames and Explosions of Gases," 2nd ed., Chapter VIII. Academic Press, New York.
von Neumann, J. (1942). OSRD Rep. No. 549.
Zeldovich, Y. B. (1950). NACA Tech. Memo. No. 1261.

Chapter Six

Diffusion Flames

Previous consideration of flames centered about fuel and air which were homogeneously distributed, i.e., premixed flames. If mixing occurs rapidly compared with combustion reactions, or well ahead of the flame zone (as in a Bunsen burner), burning may be considered in terms of homogeneous processes. There are systems, however, in which mixing is slow compared with reaction rates, so that mixing controls the burning rate. Most practical systems fall in this category, and they are the so-called diffusion flames in which the fuel and oxidant come together in a reaction zone through molecular and turbulent diffusion. The fuel may be in the form of a gaseous fuel jet or a condensed phase (liquid or solid); the distinctive characteristic of a diffusion flame is that the burning (or consumption) rate is determined by the rate at which the fuel and oxidizer are brought together in proper proportions for reaction.

Between the extremes in which the chemical reaction rate on the one hand and the mixing on the other control the burning rate, there is the region in which the chemistry and mixing have similar rates and must be considered together.

A. GASEOUS FUEL JETS

In the combustion field, gaseous diffusion flames have received less attention than premixed flame (particularly in research), despite the fact that diffusion flames have greater practical application and are used more frequently. The difficulty with gaseous diffusion flames, unlike the premixed flames, is that there is no fundamental characteristic like flame velocity which can be readily measured; even initial mixture strength has no practical meaning.

1. Appearance

Only the shape of a laminar jet of fuel depends on the mixture strength, i.e., the quantity of air supplied. If an excess of air is present, the flame is a closed, elongated figure. Such flames occur when a jet of fuel is admitted into a large volume of quiescent air, or when two coaxial jets are used, the inner containing the fuel and the outer containing an excess of air. If the air supply in the outer tube is reduced below an initial mixture strength of stoichiometric, a fan-shaped underventilated flame is produced as shown in Fig. 1.

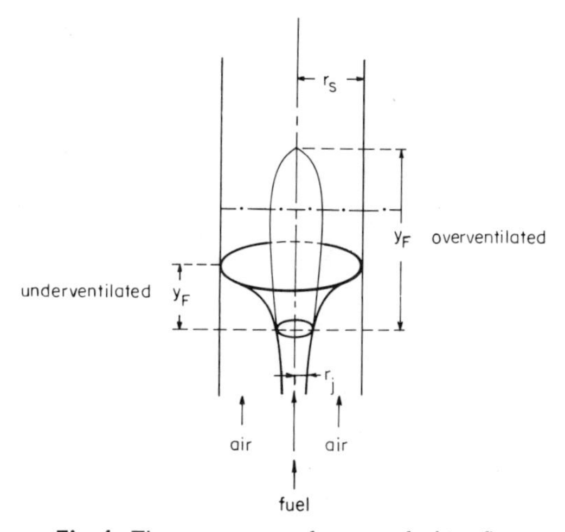

Fig. 1. The appearance of gaseous fuel jet flames.

2. Structure

Unlike the flame of premixed gases, which has a very narrow reaction zone, the diffusion flame has a wide region over which the composition of the gases changes. These changes are principally due to the interdiffusion of reactants and products since the actual reaction apparently takes place in a narrow zone.

Hottel and Hawthorne (1949) measured the distribution of species through a laminar H_2–air diffusion flame. The type of results they obtained for a radial distribution that would correspond to the broken line on Fig. 1 is shown in Fig. 2.

The concentrations of fuel and oxidizer are minima at the flame front where the product concentrations are maxima. These conditions arise, of course, due to the diffusion. The concentration profiles shown would be the same whether there were flowing fuel and oxidizer streams in the y

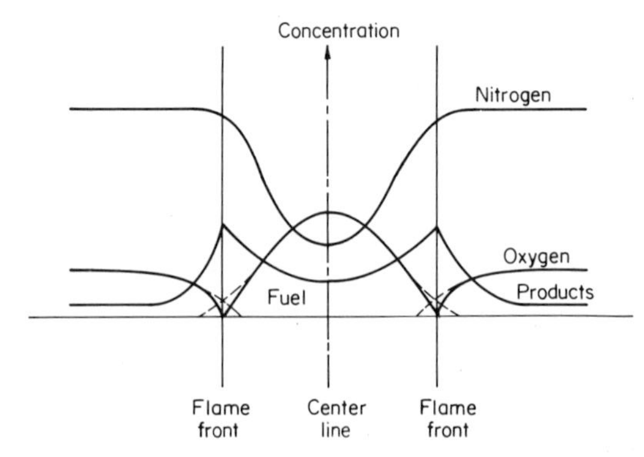

Fig. 2. Species variation through a diffusion flame at a fixed height above the fuel jet tube.

direction (as depicted in Fig. 1) or there was simply a flowing fuel jet into the infinite stagnant reservoir of the quiescent atmosphere. It is most important to realize that the diffusion establishes a bulk velocity in the direction x (or r). From the flame front outwards, there are products flowing outwards and oxygen and a little nitrogen flowing inwards. Normally in the steady state the total mass of the products is greater than the sum of the other two. Thus the bulk velocity that one would observe moves from the flame front outwards. The velocity of oxygen that arises due to the diffusion and the concentration gradient of oxygen between the outside stream and the flame front is then in the opposite direction to the bulk velocity. Between the centerline and flame front, the bulk velocity must, of course, flow from the centerline outwards. There is no sink at the centerline. In the steady state, the concentration of the products reaches a steady state value at the centerline. This value is established by the diffusion of products inward and the amount transported outward by the bulk velocity.

Since total disappearance at the flame front would indicate infinite reaction rates, it is more likely that a representation of the radial distribution should be that given by the dashed lines in Fig. 2. The dashed lines exaggerate the thickness of the flame front grossly. Even with finite rates, the flame zone is thin. Hottel and Hawthorne's experimental results show, as others do as well, that in diffusion flames the fuel and oxidizer diffuse toward each other at rates which are in stoichiometric proportions. Since the flame front is a sink for both the fuel and oxidizer, one would have intuitively expected this most important observation.

3. Theoretical Considerations

The theory of premixed flames consisted essentially of an analysis of factors such as mass diffusion, heat diffusion, and the reaction mechanisms as they affected the rate of homogeneous reactions taking place. Inasmuch as the primary mixing processes of fuel and oxidizer appear to dominate the burning processes in diffusion flames, the theories emphasize the rates of mixing (diffusion) in deriving the characteristics of such flames.

It can be verified easily by experiments that in an ethylene–oxygen premixed flame, the average rate of consumption of reactants is about 4 moles/cm^3 sec, whereas for the diffusion flame (by measurement of flow, flame height, and thickness of reaction zone—a crude but approximately correct approach) the average rate of consumption is only 6×10^{-5} moles/cm^3 sec. Thus the consumption rates in premixed flames are much larger than the pure mixing controlled diffusion flames.

The theoretical solution to the diffusion flame problem is best approached in the overall sense of a steady flowing gaseous system in which both the diffusion and chemical processes play a role. Even in burning of liquid droplets, a flow fuel due to evaporation exists. The approach is very much the same as that presented in Section 4.A.2, except that one must realize that the fuel and oxidizer are diffusing in opposite directions and in stoichiometric proportions relative to each other. If one picks a differential element along the x direction in Fig. 2, then the conservation balances for heat and mass may be obtained for the fluxes, as shown in Fig. 3. In Fig. 3, j is the mass flux as given by a representation of Fick's law of diffusion, q the heat flux, m_A the mass fraction of a species A, $\dot{m}_A$ the rate of mass generation A in the volumetric element $(\Delta x \cdot 1)$, and $\dot{H}$ the rate of heat release per unit mass of the species which is the fuel.

The general expression for the conservation of a species A (say the oxidizer) is

$$\partial \rho_A / \partial t = (\partial / \partial x)[(D\rho)\partial m_A / \partial x] - (\partial(\rho_A v)/\partial x) + \dot{m}_A \tag{1}$$

where ρ is the total mass density, ρ_A the partial density of species A, D the mass diffusivity, and v the bulk gas velocity in the direction x. The

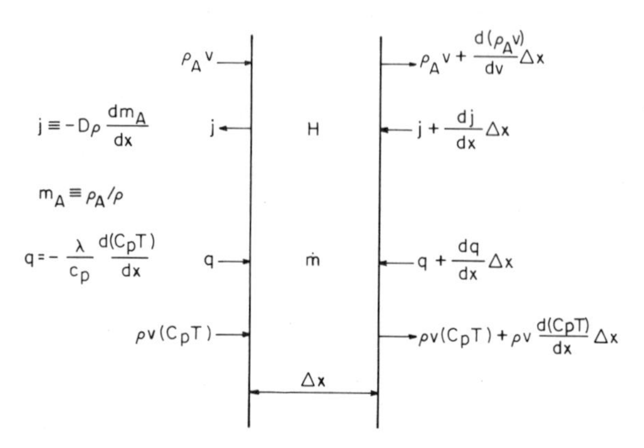

Fig. 3. Energy balance across a differential element in a diffusion flame.

time-dependent problem is outside the scope of consideration here. Note also from the definitions given in Fig. 3 that

$$\partial[(\rho_A v)]/\partial x = d[(\rho v)(\rho_A/\rho)]/dx = \rho v \, dm_A/dx \tag{2}$$

For the steady problem, with the use of Eq. (2) and the result from kinetic theory that $D\rho$ is independent of temperature, Eq. (1) becomes

$$D\rho \, d^2 m_A/dx^2 - (\rho v) \, dm_A/dx = -\dot{m}_A \tag{3}$$

Obviously the same type of expression must hold for the other diffusing species B (say the fuel) so that

$$D\rho \, d^2 m_B/dx^2 - (\rho v) \, dm_B/dx = -\dot{m}_B = -\dot{m}_A/i \tag{4}$$

where i is the mass stoichiometric coefficient $(\dot{m}_A/\dot{m}_B)$.

The energy equation evolves as it did in Section 4.A.2 to give

$$\frac{\lambda}{c_p}\frac{d^2(c_p T)}{dx^2} - (\rho v)\frac{d(c_p T)}{dx} = -\dot{H} = \left(\frac{\dot{m}_A}{i} H\right) \tag{5}$$

where $\dot{H}$ is the rate of energy release per unit mass of fuel reacted and H is the heat release per unit mass of fuel, i.e.,

$$-\dot{m}_B H = \dot{H}, \qquad -(\dot{m}_A/i)H = \dot{H} \tag{6}$$

Although the form of Eqs. (3), (4), and (5) is the same as that obtained in dealing with premixed flames, there is an important difference in the boundary conditions which exist.

By assuming Le = 1 or $D\rho = \lambda/c_p$, Eqs. (3) and (5) can be combined to give

$$D\rho(d^2/dx^2)(c_p T + m_A H/i) - (\rho v)(d/dx)(c_p T + m_A H/i) = 0 \tag{7}$$

This procedure is sometimes referred to as the Schvab–Zeldovich formulation. Mathematically what has been accomplished is that the nonhomogeneous terms ($\dot{m}$ and $\dot{H}$) have been eliminated and a homogeneous differential equation (Eq. (7)) has been obtained.

The equations could have been developed for a generalized coordinate system. In a generalized coordinate system, they would have the form

$$\nabla \cdot [(\rho v)(c_p T) - (\lambda/c_p)\nabla(c_p T)] = -\dot{H} \tag{8}$$

$$\nabla \cdot [(\rho v)(m_j) - \rho D \nabla m_j] = -\dot{m}_j \tag{9}$$

These equations could be generalized even further (see Williams, 1965) by simply writing $\sum h_j{}^0 \dot{m}_j$ instead of $\dot{H}$, where $h_j{}^0$ is the heat of formation per unit mass at the base temperature of each species j. However, for notation simplicity and since for most combustion and propulsion systems it is the energy release which is of importance, an overall rate expression for a reaction of the type which follows will suffice:

$$F + \phi O = P \tag{10}$$

where F is the fuel, O the oxidizer, P the product, and ϕ the molar stoichiometric index. Then Eqs. (8) and (9) may be written as

$$\nabla \cdot \left[(\rho v)\frac{\dot{m}_j}{\mathrm{MW}_j \nu_j} - (\rho D)\nabla \frac{\dot{m}_j}{\mathrm{MW}_j \nu_j}\right] = \dot{M} \tag{11}$$

$$\nabla \cdot \left[(\rho v)\frac{c_p T}{H\mathrm{MW}_j \nu_j} - (\rho D)\nabla \frac{c_p T}{H\mathrm{MW}_j \nu_j}\right] = \dot{M} \tag{12}$$

where MW is the molecular weight, $\dot{M} = \dot{m}_j/\mathrm{MW}_j \nu_j$; $\nu_j = \phi$ for the oxidizer, and $\nu_j = 1$ for the fuel. Both equations have the form

$$\nabla \cdot [(\rho v)\alpha - (\rho D)\nabla\alpha] = \dot{M} \tag{13}$$

where $\alpha_T = c_p T/H\mathrm{MW}_j \nu_j$ and $\alpha_j = \dot{m}_j/\mathrm{MW}_j \nu_j$. They may be expressed as

$$L(\alpha) = \dot{M} \tag{14}$$

where the linear operator $L(\ )$ is defined as

$$L(\alpha) = \nabla \cdot [(\rho v)\alpha - (\rho D)\nabla\alpha] \tag{15}$$

The nonlinear term may be eliminated from all except one of the relationships $L(\alpha) = \dot{M}$. For example,

$$L(\alpha_1) = \dot{M} \tag{16}$$

can be solved for α_1, then the other flow variables can be determined from the linear equations for a simple coupling function Ω so that

$$L(\Omega) = 0 \tag{17}$$

where $\Omega = (\alpha_T - \alpha_1) \equiv \Omega_T$ or $\Omega = (\alpha_J - \alpha_1) \equiv \Omega_j (j \neq 1)$. Obviously if 1 = fuel and there is a fuel–oxidizer system only, $j = 1$ gives $\Omega = 0$ and shows the necessary redundancy.

a. The Burke–Schumann Development

With the development in the previous section, it is now possible to approach the classical problem of the burning of a fuel jet in a coaxial stream as first described by Burke and Schumann (1928).

This description is given by the following particular assumptions:

1. At the port position, the velocities of the air and fuel are considered constant, equal, and uniform across their respective tubes. Experimentally this condition could be obtained by varying the radii of the tubes (see Fig. 1). The molar fuel rate is then given by the radii ratio:

$$r_{\mathrm{j}}^2/(r_{\mathrm{s}}^2 - r_{\mathrm{j}}^2)$$

2. The velocity of the fuel and air up the tube in the region of the flame is the same as the velocity at the port.
3. The coefficient of interdiffusion of the two gas streams is constant.

Burke and Schumann suggested that the effects of assumptions 2 and 3 compensate for each other and thus minimize errors. However, D increases as $T^{1.75}$ and velocity increases as $T^{1.00}$, but this disparity should not be the main objection. The main objection should be the variation of D with T in the horizontal direction due to heat conduction from the flame.

4. Interdiffusion is entirely radial.
5. Mixing is by diffusion only; i.e., there are no radial velocity components.
6. Of course, the general stoichiometry relation prevails.

With these assumptions it is possible to readily solve the coaxial jet problem. The only differential equation that one is obliged to consider is

$$L(\Omega) = 0 \qquad \text{with} \quad \Omega = \alpha_{\mathrm{F}} - \alpha_{\mathrm{O}}$$

where $\alpha_{\mathrm{F}} = +m_{\mathrm{F}}/\mathrm{MW_f}\, \nu_{\mathrm{F}}$ and $\alpha_{\mathrm{O}} = +m_{\mathrm{O}}/\mathrm{MW_O}\, \nu_{\mathrm{O}}$.

In cylindrical coordinates the general equation becomes

$$(v/D)(\partial\Omega/\partial y) - (1/r)(\partial/\partial r)(r\, \partial\Omega/\partial r) = 0 \tag{18}$$

The terms in $\partial/\partial\theta$ are set equal to zero because of the symmetry.

The boundary conditions become

$$\Omega = \frac{m_{F,0}}{MW_F \nu_F} \qquad \text{at} \quad y = 0, \quad 0 \leq r \leq r_j$$

$$= -\frac{m_{0,0}}{MW_0 \nu_0} \qquad \text{at} \quad y = 0, \quad r_j \leq r \leq r_s$$

and $\partial\Omega/\partial r = 0$ at $r = r_s$, $y > 0$.

It is convenient to introduce dimensionless coordinates

$$\xi \equiv r/r_s, \qquad \eta \equiv yD/vr_s^2$$

and to define parameters $c \equiv r_j/r_s$ and

$$\nu \equiv m_{0,0} MW_F \nu_F / m_{F,0} MW_0 \nu_0$$

and the reduced variable

$$\gamma \equiv \Omega(MW_F \nu_F / m_{f,0})$$

Equation (18) and the boundary condition then become

$$\partial\gamma/\partial\eta = (1/\xi)(\partial/\partial\xi)(\xi\, \partial\gamma/\partial\xi) \tag{19}$$

$$\gamma = 1 \quad \text{at} \quad \eta = 0, \quad 0 \leq \xi < c; \qquad \gamma = -\nu \quad \text{at} \quad \eta = 0, \quad c \leq \xi < 1$$

and

$$\partial\gamma/\partial\xi = 0 \qquad \text{at} \quad \xi = 1, \quad \eta > 0$$

Equation (19) with these new boundary conditions has the known solution:

$$\gamma = (1 + \nu)c^2 - \nu + 2(1 + \nu)c \sum_{n=1}^{\infty} (1/\phi_n)(J_1(c\phi_2)/[J_0(\phi_n)^2]) \; J_0(\phi_n \xi) \exp(-\phi_n^2) \tag{20}$$

where J_0 and J_1 are Bessel functions of the first kind (of order zero and one, respectively) and the ϕ_n represent successive roots of the equation $J_1(\phi) = 0$ (with ordering convention $\phi_n > \phi_{n-1}$, $\phi_0 = 0$). This equation gives the solution for Ω in the present problem.

The flame shape is defined at the point where the fuel and oxidizer disappear and that specifies the place where $\Omega = 0$. Hence, setting $\gamma = 0$ provides a relation between ξ and η that defines the locus of the flame surface.

The equation for the flame height is obtained by solving Eq. (20) for η after setting $\xi = 0$ for the overventilated flame and $\xi = 1$ for the underventilated flame (also $\gamma = 0$).

The resulting equation is still very complex. Since flame heights are large enough to cause the factor $\exp(-\phi_n{}^2\eta)$ to decrease rapidly as n increases at these values of η, it usually suffices to retain only the first few terms of the sum in the basic equation for this calculation. Neglecting all terms except $n = 1$, one obtains the rough approximation

$$\eta = (1/\phi_1{}^2) \ln\{2(1 + \nu)cJ_1(c\phi_1)/[\nu - (1 + \nu)c]\phi_1 J_0(\phi_1)\} \qquad (21)$$

for the dimensionless height of the overventilated flame. The first zero of $J_1(\phi)$ is $\phi_1 = 3.83$.

The flame shapes and heights predicted by such expressions (see Fig. 4) are shown by Lewis and von Elbe (1961) to be in good agreement with

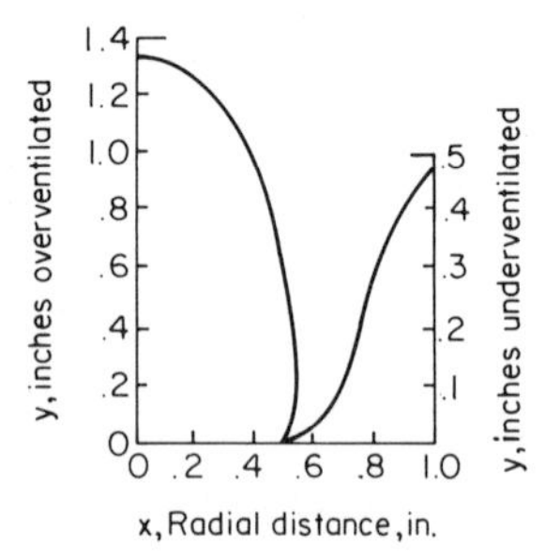

Fig. 4. Flame shapes as predicted by Burke and Schuman for cylindrical fuel jet systems (after Lewis and von Elbe, 1961).

experimental results—which is surprising considering the basic drastic assumptions.

b. Phenomenological Analysis and the Turbulent Fuel Jet

Actually much information can be gleaned from some simple phenomenological reasoning applied to the burning of a free fuel jet. The fundamental assumption which has been made is that the combustion process does not affect the mixing rate between the fuel jet and the surrounding oxidizer. As soon as the oxidizer mixes into the fuel it reacts. In the cylindrical problem, as soon as the oxidizer has diffused from the jet edge to the centerline, the fuel must have completely burned; in other words, the flame ends at this point and the flame height is known.

The problem, then, is simply one of laminar (diffusional) mixing of a jet—which in itself can be approached phenomenologically.

Simple kinetic theory predicts that in molecular diffusion the average displacement of a molecule is given by

$$\tfrac{1}{2} d\overline{X^2}/dt = D$$

where D is the diffusion coefficient and $\overline{X^2}$ is the mean square of the displacement of the particles in a specific direction in the time t. Thus to arrive at an estimate of the flame length, it is proposed to use this equation from kinetic theory in the integrated fashion

$$\xi^2 = 2Dt$$

where ξ^2 denotes the average of the squares of displacement of molecules from some location due to diffusion during time t. The length of the flame is assumed to correspond to the condition that at the point on the stream axis where combustion is complete, the average depth of penetration of air into fuel must be approximately equal to the radius (proportional to radius more exactly) of the burner tube. As an approximation, ξ is identified with the average depth of penetration. The gas velocity u is taken as constant, so that the time t, the time required for completion of the diffusion process—which is the time a gas element flows from the burner port to the flame tip—is given by

$$t = y_F/u$$

It follows since $\xi = r_j$, that

$$y_{F,L} = r_j^2 u/2D = Q/2\pi D \tag{22}$$

where Q is the volumetric flowrate. This result is significant because not only does it predict that for the laminar diffusion flame the height is proportional only to the volumetric flowrate, but it gives us the best answer for the very important problem of turbulent fuel jets.

If one examines Eq. (21) and the definition of η in the Burke–Schumann approach, it becomes apparent that elements of the phenomenological result represented by Eq. (22) are indeed in the Burke–Schumann results.

For the turbulent case, the same reasoning can be used except that instead of the molecular diffusivity the turbulent eddy diffusivity ε is used and a proportionality similar to Eq. (22) is developed,

$$y_{F,T} \sim \pi r_j^2 u/2\pi\varepsilon \sim r_j^2 u/\varepsilon \tag{23}$$

But $\varepsilon \sim lu'$, where l is the scale of the turbulence and is proportional to the tube diameter (or tube radius) and u' is the intensity of turbulence, which is approximately proportional to the mean flow velocity at the axis. Or

$$\varepsilon \sim r_j u \tag{24}$$

Combining the last two equations, one obtains

$$y_{F,T} \sim r_j^2/r_j \sim r_j \tag{25}$$

This expression shows that the height of a turbulent diffusion flame is proportional to the port radius (or diameter) only!! This is a very important practical conclusion that has been verified in many ways. Figure 5 shows the variation of flame versus the tube velocity as characterized by Hottell and Hawthorne (1949).

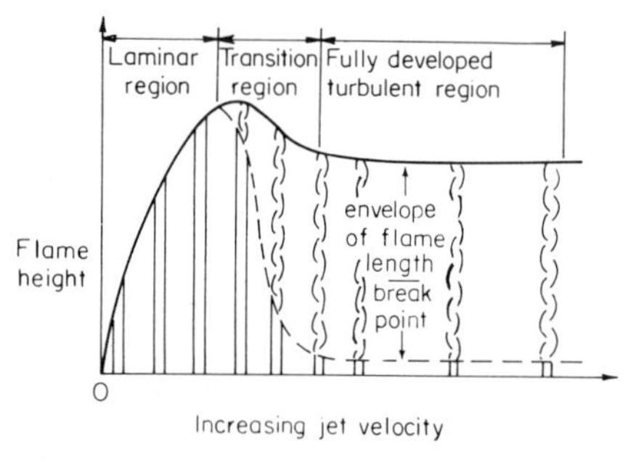

Fig. 5. The variation of flame character as a function of jet velocity (after Hottel and Hawthorne, 1949).

B. BURNING OF CONDENSED PHASES

When most liquids or solids are projected into an atmosphere so that a combustible mixture is formed and when this mixture is ignited, a flame surrounds the liquid or solid phase. Except at the very lowest of pressures, around 10^{-6} Torr, this flame is a diffusion flame. If the condensed phase is considered as a liquid fuel and the gaseous oxidizer as oxygen, then fuel is evaporated from the liquid interface and diffuses to the flame front as the oxygen moves from the surroundings to the burning front. This picture of condensed phase burning is most readily and usually applied to droplet burning but can also be applied to any liquid surface.

The rate at which the droplet evaporates and burns is generally considered to be determined by the rate of heat transfer from the flame front to the fuel surface. Here, as in the case of gaseous diffusion flames, chemical processes are assumed to occur so rapidly that the burning rates are determined solely by mass and heat transfer rates.

Most of the analytical models of this burning process consider a double-film model for the combustion of the liquid fuel. One film separates the droplet surface from the flame front and the other separates the flame front from the surrounding oxidizer atmosphere as depicted in Fig. 6.

In most analytical developments the liquid surface is assumed to be at the normal boiling point temperature of the fuel. Surveys of the temperature field by Khudyakov (1955) in burning liquids indicate that the

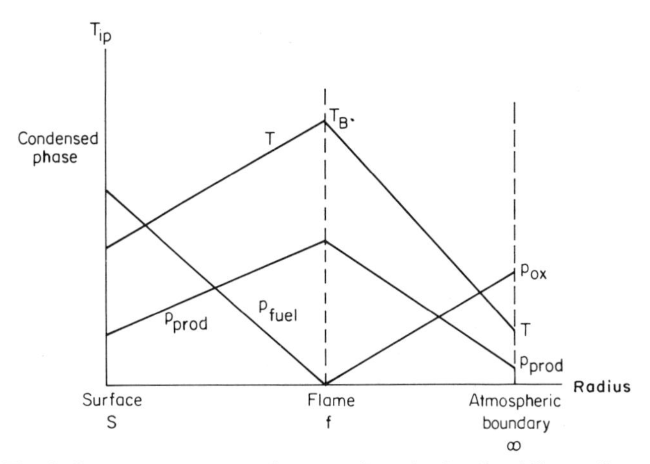

Fig. 6. Parameter variation along a radius of a droplet diffusion flame.

temperature is only a few degrees below the boiling temperature. In the approach to be employed here, all that is required is that the droplet be at a uniform temperature at or below the normal boiling point. In the sf region of Fig. 6 fuel evaporates at the drop surface and diffuses towards the flame front where it is consumed. Heat is conducted from the flame front to the liquid and it vaporizes the fuel. Most analyses assume that the fuel is heated to the flame temperature before it chemically reacts and that the fuel does not react until it reaches the flame front. This latter assumption means that the flame front is a mathematically thin surface where the fuel and oxidizer meet in stoichiometric proportions. Most early investigators did not realize that it was not necessary to determine T_f in order to calculate the fuel burning rate. However, in order to determine a T_f, the infinitely thin reaction zone at the stoichiometric position must be assumed.

In the film $f\infty$, oxygen diffuses to the flame front and combustion products and heat are transported to the surrounding atmosphere. The position of the boundary designated by ∞ is determined by convection. A stagnant atmosphere places the boundary at an infinite distance from the fuel surface.

It should be stated that all analyses assume no radiant energy transfer.

1. General Mass Burning Considerations and the Evaporation Coefficient

Three parameters are generally evaluated: the mass burning rate, the flame position above the fuel surface, and the flame temperature. The most important parameter is the mass burning rate, for it permits the evaluation of the so-called evaporation coefficient, which is most readily measured experimentally.

The use of the term evaporation coefficient comes about from mass and heat transfer experiments without combustion; i.e., evaporation, generally as used in spray drying and humidification. Basically the evaporation coefficient β is defined by the following expression which has been verified experimentally:

$$d^2 = d_0{}^2 - \beta t \tag{26}$$

where d_0 is the original drop diameter and d is the drop diameter after the time t. It will be shown later that the same expression has been found to hold for mass and heat transfer with chemical reaction.

It should be realized that the combustion of droplets is one aspect of a much broader problem which involves the gasification of a condensed phase, i.e., a liquid *or a solid*. In this sense, the field of diffusion flames is rightfully broken down into gases and condensed phases. Here the concern is with the burning of droplets, but the concepts to be used are just as applicable to the evaporation of liquids, sublimation of solids, hybrid burning rates, ablation heat transfer, solid propellant burning, transpiration cooling, etc. In all cases one is interested in the mass consumption, or the rate of regression, of a condensed material. Whereas in gaseous diffusion flames there was no specific property to measure and the flame height was evaluated, in condensed diffusion flames a specific quantity is measurable. This quantity is some representation of the mass consumption rate of the condensed phase. The similarity of the cases just mentioned arises because the condensed phase must be gasified, and consequently there must be an energy input into the condensed material. What determines the rate of regression or evolution of material is the heat flux at the surface. Thus in all the cases just mentioned,

$$q = \dot{r}\rho_f H_A \tag{27}$$

where q is the heat flux to the surface in cal/cm^{2}sec, $\dot{r}$ is the regression rate cm/sec; ρ_f is the density of the condensed phase and H_A is the overall energy per unit mass required to gasify the material. H_A is usually the sum of two terms—the heat of vaporization, sublimation, or gasification plus the enthalpy required to bring the surface to the temperature of vaporization, sublimation, or gasification.

From the foregoing discussion, it is seen that the heat flux, H_A and the density determine the regression rate, but it must also be realized that this statement does not mean the heat flux is the controlling or rate-determining step in each case. The fact is that it is generally not the controlling step.

The controlling step and the heat flux are always interrelated however. Regardless of the process of concern (assuming no radiation)

$$q = -\lambda(\partial T/\partial y)_s \tag{28}$$

where λ is the thermal conductivity and the subscript designates the surface. This simple statement of the Fourier heat conduction law is of such great significance that its importance cannot be overstated.

This same equation holds regardless if there is mass evolution from the surface and regardless if convective effects prevail. For even in convective atmospheres where one is interested in the heat transfer to surface (without mass addition of any kind, i.e., in the heat transfer situation generally encountered), one writes the heat transfer equation as

$$q = h(T_\infty - T_s) \tag{29}$$

It must be realized that this statement is shorthand for

$$q = -\lambda(\partial T/\partial y)_s = h(T_\infty - T_s) \tag{30}$$

T_∞ and T_s are the free stream and surface temperatures, respectively; the heat transfer coefficient h is by definition

$$h \equiv \lambda/\delta \tag{31}$$

where δ is the boundary layer thickness. Again by definition, the boundary layer is the distance between the surface and free stream condition, thus as an approximation

$$q = \lambda(T_\infty - T_s)/\delta \tag{32}$$

The $(T_0 - T_s)/\delta$ term is the temperature gradient which correlated $\lambda(\partial T/\partial y)_s$ through the boundary layer thickness. The fact that δ can be correlated with the Reynold's number and that the Colburn analogy can be applied is what leads to the correlations of the form

$$\mathrm{Nu} = f(\mathrm{Re}, \mathrm{Pr}) \tag{33}$$

where Nu is the Nusselt number hx/λ; Pr is the Prandtl number $c_p\mu/\lambda$; and $\mathrm{Re} = \rho vx/\mu$ the Reynold's number where x is the critical dimension—the distance from the leading edge of a flat plate or the diameter of a tube.

Although the correlations given by Eq. (33) are useful for practical evaluation of heat transfer to a wall, one must not lose sight of the fact that it is the temperature gradient at the wall that actually determines the heat flux there. In transpiration cooling problems, it is not so much that the injection of the transpiring fluid increases the boundary layer thickness and thus decreases the heat flux, rather that the temperature gradient at the surface is decreased by the heat absorbing capacity of the fluid. What

Eq. (28) specifies is that regardless of the processes taking place, the temperature profile at the surface determines the regression rate—whether it be evaporation, solid propellant burning, etc. Thus all the mathematical approaches used in the type of problems mentioned before simply seek to evaluate the temperature gradient at the surface. The temperature gradient at the surface *is different* for the various processes discussed. Thus the temperature profile from the surface to the source of energy will be different for evaporation than for the burning of a liquid fuel which releases energy when oxidized.

Nevertheless, a diffusion mechanism generally prevails, is the slowest step, and thus determines the regression rate. In evaporation, it is the conduction of heat from the surrounding atmosphere to the surface; in ablation it is the conduction of heat through the boundary layer; in droplet burning it is the diffusion rates of the fuel as it diffuses to approach the oxidizer, etc.

It is interesting from a mathematical sense to realize that the gradient at the surface will always be a boundary condition to the mathematical statement of the problem. Thus the mathematical solution is necessary simply to evaluate the boundary condition.

Further, it should be emphasized that the absence of radiation has been assumed. To incorporate radiation is not difficult if the assumption is made that the radiant intensity of the emitters is known and there is no absorption between the emitters and the vaporizing surfaces; i.e., it can be assumed that q_r the radiant heat flux to the surface is known. Then Eq. (33) becomes

$$q + q_r = -\lambda(\partial T/\partial y)_s + q_r = \dot{r}\rho H_A \qquad (34)$$

It is interesting to note that for the assumptions above the mathematical solution of the problem does not become significantly more difficult, for again q_r and thus radiation effects enter only in the boundary condition, and the differential equations describing the processes are not altered.

It is the intent to first calculate the evaporation rate of a single fuel droplet before considering the combustion of this fuel droplet or, to say it more exactly, the evaporation of a fuel droplet in the presence of combustion. Since one is concerned now with diffusional processes, it is best to start by reconsidering the laws that hold for diffusional processes.

Fick's law says that if a gradient in concentration of species A exists, say (dn_A/dy), then there is found a flow or flux of A say j_A across a unit area in the y direction such that

$$j_A = -D\, dn_A/dy \qquad (35)$$

where D is called the mass diffusion coefficient or more simply the diffusion coefficient, n_A is the number concentration of molecules (no./cm^3), and j is the flux of mass, number of molecules/cm^2 sec. Thus the units of D are cm^2/sec.

The Fourier law of heat conduction relates the flux of heat q per unit area, as a result of a temperature gradient, such that

$$q = -\lambda\, dT/dy$$

The units of q are cal/cm^2 sec and those of λ are cal/cm sec °K. It is not the temperature, an intensive property, that is exchanged, but energy. In this case, the energy density and the exchange reaction, which show similarity, are written as

$$q = -\frac{\lambda}{\rho c_p} \cdot \rho c_p \frac{dT}{dy} = -\frac{\lambda}{\rho c_p}\frac{dH}{dy} = -\alpha \frac{dH}{dy} \tag{36}$$

where α is called the thermal diffusivity and has units cm^2/sec since λ = cal/cm sec °K, c_p = cal/gm °K, and ρ = gm/cm^3. H is the energy concentration in cal/cm^3. Thus the similarity of Fick's and Fourier's laws is apparent. One arises due to a number concentration gradient and the other due to an energy concentration gradient.

A similar law to these two diffusional processes is Newton's law of viscosity which relates the flux (or shear stress) τ_{yx} of the x component of momentum u_x due to a gradient in u_x, and is written as

$$\tau_{yx} = -\mu\, du_x/dy \tag{37}$$

where the units of the stress are gm/cm^2 and of the viscosity are gm/cm sec. Again, it is not velocity which is exchanged, but momentum and when the exchange of momentum density is written, similarity is noted

$$\tau_{yx} = -(\mu/\rho)(d(\rho u_x)/dy) = -\nu(d(\rho u_x)/dy) \tag{38}$$

where ν is the momentum diffusion coefficient or more acceptably, the kinematic viscosity. Since Eq. (38) relates the momentum gradient to flux, its similarity to Eqs. (35) and (36) is seen readily. Recall, as stated earlier in Chapter 4, that simple kinetic theory of gases gives $\alpha = D = \nu$. The ratios of these three diffusivities gives some familiar dimensionless similarity parameters

$$\mathrm{Pr} = \nu/\alpha, \qquad \mathrm{Sc} = \nu/D, \qquad \mathrm{Le} = \alpha/D$$

where Pr, Sc, and Le are Prandtl, Schmidt, and Lewis numbers, respectively. Thus for gases simple kinetic theory gives as a first approximation

$$\mathrm{Pr} = \mathrm{Sc} = \mathrm{Le} = 1$$

2. Single Fuel Droplets in Quiescent Atmospheres

Since Fick's law will be used in many different forms in the ensuing development, perhaps it is best to develop those forms now so that the later developments need not be interrupted.

Consider the diffusion of molecules A into an atmosphere of B molecules, i.e., a binary system. For a concentration gradient in A molecules alone, the further developments can be simplified readily if Fick's law is now written

$$j_A = -D_{AB}\, dn_A/dy \tag{39}$$

where D_{AB} is the binary diffusion coefficient. n_A is now considered in number of moles per unit volume since one could always multiply Eq. (35) through by Advogadro's number. j_A is now expressed in number of moles as well.

Multiplying through Eq. (39) by MW_A the molecular weight of A, one obtains

$$(j_A\, MW_A) = -D_{AB}\, d(n_A\, MW_A)/dy = -D_{AB}\, d\rho_A/dy \tag{40}$$

ρ_A is the mass density of A, ρ_B mass density of B, and ρ = total density = $\rho_A + \rho_B$ with units gm/cm^3.

If there is no mole accumulation or creation, then the total number of moles is constant

$$n = n_A + n_B \tag{41}$$

and

$$dn/dy = 0, \qquad dn_A/dy = -dn_B/dy, \qquad j_A = -j_B \tag{42}$$

Thus there will result a net flux of mass by diffusion equal to

$$\rho v = j_A\, MW_A + j_B\, MW_B = j_A(MW_A - MW_B) \tag{43}$$

where v is the bulk direction velocity established by the diffusional masses.

In problems dealing with the combustion of condensed matter and thus regressing surfaces, there is always a bulk velocity movement in the gases. Thus the situation exists in which there are species diffusing while the bulk gases are moving at a finite velocity. The diffusion of the species can be against or with the bulk flow (velocity). For mathematical convenience it is best to decompose flows in which there is diffusion into a flow with an average mass velocity v and a diffusive velocity relative to v.

When one gas diffuses into another, as A into B, even without the quasi-steady flow component imposed by the burning, the mass transport of a species, say A, is made up of two components—the normal diffusion component and the component related to the bulk movement established by the diffusion process. This mass transport flow has a velocity Δ_A and the mass of A transported per unit area is $\rho_A \Delta_A$. The bulk velocity established by the diffusive flow is given by Eq. (43). The fraction of that flow is Eq. (43) multiplied by the mass fraction of A, ρ_A/ρ. Thus

$$\rho_A \Delta_A = j_A \text{MW}_A - j_A(\text{MW}_A - \text{MW}_B)\rho_A/\rho \tag{44}$$

where the first term on the right-hand side is the normal diffusive flux in mass units and the second term is the bulk movement flow due to diffusion, as just described. Writing the flux j in terms of the mass diffusivity,

$$\rho_A \Delta_A = -D_{AB}(d\rho_A/dy)[1 - (1 - \text{MW}_B/\text{MW}_A)(\rho_A/\rho)] \tag{45}$$

Since $j_A = -j_B$

$$\text{MW}_A j_A = -\text{MW}_A j_B = -(\text{MW}_B j_B)(\text{MW}_A/\text{MW}_B) \tag{46}$$

and

$$-d\rho_A/dy = (\text{MW}_A/\text{MW}_B)(d\rho_B/dy) \tag{47}$$

However

$$(d\rho_A/dy) + (d\rho_B/dy) = d\rho/dy \tag{48}$$

which gives with Eq. (47)

$$(d\rho_A/dy) - (\text{MW}_B/\text{MW}_A)(d\rho_A/dy) = d\rho/dy \tag{49}$$

Multiplying through by ρ_A/ρ, one obtains

$$(\rho_A/\rho)[1 - (\text{MW}_B/\text{MW}_A)](d\rho_A/dy) = (\rho_A/\rho)(d\rho/dy) \tag{50}$$

Substituting Eq. (50) into Eq. (45), one obtains

$$\rho_A \Delta_A = -D_{AB}((d\rho_A/dy) - (\rho_A/\rho)(d\rho/dy)) \tag{51}$$

or

$$\rho_A \Delta_A = -\rho D_{AB}\, d(\rho_A/\rho)/dy \tag{52}$$

Defining m_A as the mass fraction of A,

$$\rho_A \Delta_A = -\rho D_{AB}\, dm_A/dy \tag{53}$$

The total mass flux of A under the condition of the burning of a condensed phase, which imposes a bulk velocity related to the mass burned, is then

$$\rho_A v_A = \rho_A v + \rho_A \Delta_A = \rho_A v - \rho D_{AB}\, dm_A/dy \tag{54}$$

$\rho_A v$ is the bulk transport part and $\rho_A \Delta_A$ is the diffusive transport part. Indeed, in the developments of Chapter 5, the correct diffusion term was used without the developments just completed.

a. Heat and Mass Transfer without Chemical Reaction (Evaporation)—the Transfer Number B

Following Blackshear's (1960) adaptation of Spalding's approach (1953, 1955), consideration will now be given to the calculation of the evaporation of a single fuel droplet in a nonconvective atmosphere of a given temperature and pressure. A major assumption is now made in that the problem is considered as a quasi-steady one. Thus the droplet is of fixed size and retains that size by a steady flux of fuel. One can consider the regression as being instant or even better to think of the droplet as a porous sphere being fed from a very thin tube at a rate equal to the mass burning so that the surface of the sphere is always wet and immediately replaces any liquid evaporated. The porous sphere approach shows that for the diffusion flame a bulk gaseous velocity outward must exist and that this velocity in the spherical geometry will vary but must always be the value given by $m = 4\pi r^2 \rho v$. This velocity is the one referred to in the last section. With this physical picture, the important assumption is made that the temperature throughout the droplet is constant and equal to the surface temperature.

In developing the problem, a differential volume in the vapor above the liquid droplet is chosen as shown in Fig. 7. The surface area of the sphere is $4\pi r^2$. Since mass cannot accumulate in the element,

$$d(\rho A v) = 0, \qquad (d/dr)(4\pi r^2 \rho v) = 0 \tag{55}$$

which is essentially the continuity equation.

Consider now the energy equation of the evaporating drop in spherical-symmetric coordinates in which c_p and λ are taken independent of temperature. The heat entering at the surface (i.e., the amount of heat convected in) is $\dot{m} c_p T$ or $(4\pi r^2 \rho v) c_p T$ (see Fig. 7). The heat leaving after

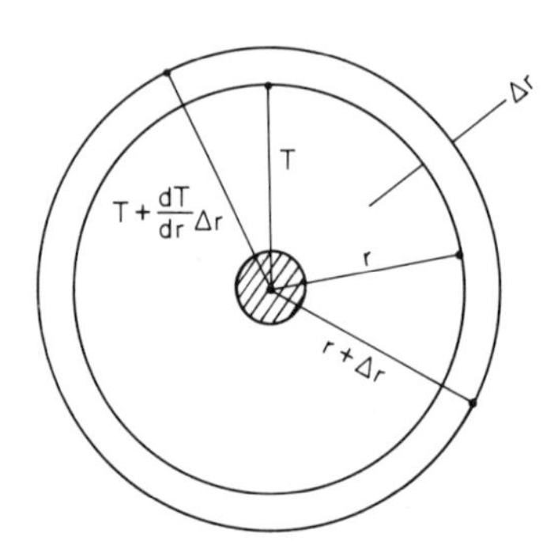

Fig. 7. The temperature balance across a differential element in spherical symmetry.

$r + \Delta r$ is $\dot{m}c_p(T + (dT/dr)\,\Delta r)$ or $(4\pi r^2 \rho v)c_p(T + (dT/dr)\,\Delta r)$. The difference then is

$$4\pi r^2 \rho v c_p (dT/dr)\,\Delta r$$

The heat diffusing from r toward the drop (out of the element) is

$$-\lambda 4\pi r^2\, dT/dr$$

The heat diffusing into the element is

$$-\lambda 4\pi (r + \Delta r)^2 (d/dr)[T + (dT/dr)\,\Delta r]$$

or

$$-[\lambda 4\pi r^2 (dT/dr) + \lambda 4\pi r^2 (d^2T/dr^2)\,\Delta r] + \lambda 8\pi r^2\,\Delta r\, dT/dr$$

$$r + \text{two terms in } \Delta r^2 \text{ and one in } \Delta r^3 \text{ which are negligible}$$

The difference in the two terms is

$$-[\lambda 4\pi r^2 (d^2T/dr^2)\,\Delta r + 8\lambda\pi r(dT/dr)]$$

Heat could be generated in the volume element defined by Δr and one has

$$(4\pi r^2\,\Delta r)\dot{H}$$

where $\dot{H}$ is the heat release rate per unit mass in the volumetric element. Thus for the energy balance

$$4\pi r^2 \rho v c_p (dT/dr)\,\Delta r - [\lambda 4\pi r^2 (d^2T/dr^2)\,\Delta r + \lambda 8\pi r\,\Delta r(dT/dr)]$$
$$= 4\pi r^2 \cdot \Delta r \dot{H} \tag{56}$$

$$4\pi r^2 \rho v c_p (dT/dr) = \lambda 4\pi r^2 (d^2T/dr^2) + 2\lambda 4\pi r^2 (dT/dr) + 4\pi r^2 \dot{H} \tag{57}$$

or

$$4\pi r^2 (dc_p\,T/dr) = d/dr[(\lambda 4\pi r^2/c_p)(dc_p\,T/dr)] + 4\pi r^2 \dot{H} \tag{58}$$

Similarly the conservation of the Ath species can be written as

$$4\pi r^2 \rho v\, dm_A/dr = (d/dr)(4\pi r^2 \rho D\, dm_A/dr) + 4\pi r^2 \dot{m}_A \tag{59}$$

where $\dot{m}_A$ is the generation rate of A due to reaction in the unit volume. The kinetic theory of gases gives to a first approximation that the product $D\rho$ (and thus, λ/c_p) is independent of temperature and pressure, consequently $D_s\rho_s = D\rho$.

Consider a droplet of radius r. If the droplet is vaporizing, then the fluid will leave the surface by convection and diffusion. Since at the droplet surface only A exists, then the boundary condition at the surface is

$$\underbrace{\rho_s v_s}_{\text{amount of material leaving the surface}} = \rho m_{As} v_s - \rho D(dm_A/dr)_s \tag{60}$$

Equation (60) is, of course, explicitly Eq. (54). $\rho_s v_s$ is the bulk mass movement which at the surface is exactly the amount of A which is being convected plus the amount of gaseous A which diffuses to or from the surface. Equation (60), then, is written as

$$v_s = D(dm_A/dr)_s/(m_{As} - 1)$$

In the sense of Spalding, a new parameter is defined

$$b \equiv m_A/(m_{As} - 1) \tag{61}$$

Equation (61) thus becomes

$$v_s = D(db/dr)_s \tag{62}$$

and Eq. (59) becomes

$$r^2 \rho v \, db/dr = (d/dr)(r^2 \rho D \, db/dr) \tag{63}$$

The boundary condition at $r = \infty$ is $m_A = m_{A\infty}$ or

$$b = b_\infty \qquad \text{at} \quad r \to \infty \tag{64}$$

From continuity

$$r^2 \rho v = r_s^2 \rho_s v_s \tag{65}$$

Integration of Eq. (63) proceeds simply since $r^2 \rho v =$ constant on the left-hand side of this equation

$$r^2 \rho v b = r^2 \rho D \, db/dr + \text{const} \tag{66}$$

Evaluating the constant at $r = r_s$, one obtains

$$r_s^2 \rho_s v_s b_s = r_s^2 \rho_s v_s + \text{const}$$

since from Eq. (62) $v_s = D(db/dr)_s$. Or, one has for Eq. (66)

$$r_s^2 \rho_s v_s (b - b_s + 1) = r^2 \rho D \, db/dr \tag{67}$$

By separating variables,

$$(r_s^2 \rho_s v_s / r^2 \rho D) \, dr = db/(b - b_s + 1) \tag{68}$$

assuming ρD constant and integrating (recall $\rho D = \rho_s D_s$)

$$-(r_s^2 v_s / r D_s) = \ln(b - b_s + 1) + \text{const} \tag{69}$$

Evaluating the constant at $r \to \infty$, one obtains

$$\text{const} = -\ln(b_\infty - b_s + 1)$$

or Eq. (69) becomes

$$r_s^2 v_s / r D_s = \ln\{(b_\infty - b_s + 1)/(b - b_s + 1)\} \tag{70}$$

The left-hand term of Eq. (69) goes to zero as $r \to \infty$. This point is significant because it shows that the spherical–symmetric, quiescent case is the only mathematical tractable quiescent case. No other quiescent case such as that for cylindrical symmetry or any other symmetry is tractable. Evaluating Eq. (70) at $r = r_s$ results in

$$\frac{r_s v_s}{D_s} = \ln(b_\infty - b_s + 1) = \ln(1 + B)$$
$$r_s v_s = D_s \ln(b_\infty - b_s + 1) = D_s \ln(1 + B) \tag{71}$$

where $B \equiv b_\infty - b_s$. The mass flowrate per unit area $G_A = \dot{m}_A/4\pi r_s^2$, where $\dot{m}_A = \rho_s v_s 4\pi r_s^2$, then

$$4\pi r_s \rho_s r_s v_s / 4\pi r_s \rho_s D_s = \ln(1 + B)$$
$$G_A = \dot{m}_A/4\pi r_s^2 = (\rho_s D_s/r_s) \ln(1 + B) = (D\rho/r_s) \ln(1 + B) \tag{72}$$

Since the product $D\rho$ is independent of pressure, the evaporation rate is also independent of pressure. In order to find a solution for Eq. (72), or more rightly to evaluate B, m_{As} must be determined. A reasonable assumption would be that the gas which surrounds the droplet is saturated at the surface temperature. Thus the problem now is to determine T_s since vapor pressure data are available.

For the case of evaporation, Eq. (58) becomes

$$r_s^2 \rho_s v_s c_p \, dT/dr = (d/dr)(r^2 \lambda \, dT/dr) \tag{73}$$

Integrating,

$$r_s^2 \rho_s v_s c_p T = r^2 \lambda \, dT/dr + \text{const} \tag{74}$$

The boundary condition at the surface is

$$(\lambda \, dT/\partial r)_s = \rho_s v_s L_v \tag{75}$$

where L_v is the latent heat of vaporization at the temperature T_s. Recall that the droplet is considered uniform throughout at the temperature T_s. Substituting Eq. (75) into Eq. (74)

$$\text{const} = r_s^2 \rho_s v_s [c_p T_s - L_v]$$

Thus Eq. (74) becomes

$$r_s^2 \rho_s v_s c_p [T - T_s + L_v/c_p] = r^2 \lambda \, dT/dr \tag{76}$$

Integrating Eq. (76),

$$-r_s^2 \rho_s v_s c_p / r\lambda = \ln[T - T_s + (L_v/c_p)] + \text{const} \tag{77}$$

After evaluating the constant at $r \to \infty$, one obtains

$$\frac{r_s^2 \rho_s v_s c_p}{r\lambda} = \ln\left[\frac{T_\infty - T_s + (L_v/c_p)}{T - T_s + (L_v/c_p)}\right] \tag{78}$$

Evaluating Eq. (78) at the surface,

$$\frac{r_s \rho_s v_s c_p}{\lambda} = \ln\left(\frac{c_p(T_\infty - T_s)}{L_v} + 1\right) \tag{79}$$

Since $\alpha = \lambda / c_p \rho$,

$$r_s v_s = \alpha_s \ln\left(1 + \frac{c_p(T_\infty - T_s)}{L_v}\right) \tag{80}$$

Comparing Eqs. (71) and (80), one can write

$$r_s v_s = \alpha_s \ln\left(\frac{c_p(T_\infty - T_s)}{L_v} + 1\right) = D_s\left(\frac{m_{A\infty} - m_{As}}{m_{As} - 1} + 1\right)$$

or

$$r_s v_s = \alpha_s \ln(1 + B_T) = D_s \ln(1 + B_M) \tag{81}$$

where

$$B_T \equiv \frac{c_p(T_\infty - T_s)}{L_v}, \qquad B_M \equiv \frac{m_{A\infty} - m_{As}}{m_{As} - 1}$$

Again since $\alpha = D$,

$$B_T = B_M$$

and

$$c_p(T_\infty - T_s)/L_v = (m_{A\infty} - m_{As})/(m_{As} - 1) \tag{82}$$

m_{As} is determined from the vapor pressure of A or the fuel; however, it must be realized that m_{As} is a function of the total pressure since

$$m_{As} \equiv \rho_{As}/\rho = n_A \mathrm{MW}_A / n\mathrm{MW} = (P_A/P)(\mathrm{MW}_A/\mathrm{MW}) \tag{83}$$

where n_A and n are the number of moles of A and the total number of moles, respectively, MW_A and MW are the molecular weight of A and the average molecular weight of the mixture, respectively, and P_A and P are the vapor pressure of A and the total pressure, respectively.

In order to obtain the solution desired, a value of T_s is assumed, the vapor pressure of A determined from tables, and m_{As} calculated from Eq. (83). This value of m_{As} and the assumed value of T_s are inserted in Eq. (82). If this equation is satisfied, then the correct T_s was chosen. If not then one must reiterate. When the correct value of T_s and m_{As} are found, then B_T or B_M are determined for the given initial conditions T_∞ or $m_{A\infty}$. For fuel combustion problems, $m_{A\infty}$ is usually zero, however, for evaporation, say of water, there is humidity in the atmosphere and this humidity must be represented as $m_{A\infty}$. With B_T and B_M determined, the mass evaporation rate is determined from Eq. (72) for a fixed droplet size. It is, of course, much more preferable to know the evaporation coefficient β from which the total evaporation time can be determined. Once B is known, the evaporation coefficient can be determined readily as will be shown later.

b. Heat and Mass Transfer with Chemical Reaction (Burning Rates)

The previous development can also be used to determine the burning rate, or evaporation coefficient, of a single droplet of fuel in a quiescent atmosphere. In this case, the mass fraction of the fuel which is always the condensed phase will be designated m_f, the mass fraction of the oxidizer m_0 the oxidant exclusive of inerts, and i the mass stoichiometric fuel–oxidant ratio exclusive of inerts. The same assumptions as those made for evaporation from a porous sphere hold. Recall, as well, that the temperature throughout the droplet is the same and is assumed to be equal to the surface temperature T_s. For liquid fuels, this temperature is generally very near, but slightly less, than the saturation temperature.

Again, it is assumed for combustion that the fuel and oxidant approach each other in stoichiometric proportions. Thus, as before, the following relationships can be written

$$\dot{m}_f = \dot{m}_0 i, \qquad \dot{m}_f H = \dot{m}_0 H i = -\dot{H}$$

where H is the heat of reaction of the fuel per unit mass.

If one defines the variables

$$\begin{aligned} b_{f0} &\equiv (m_f - i m_0)/((m_{fs} - 1) + (m_0)_s i) \\ b_{fq} &\equiv (m_f H + c_p T)/(L_v + H(m_{fs} - 1)) \\ b_{0q} &\equiv (m_0 i H + c_p T)/(L_v + H i (m_0)_s) \end{aligned} \tag{84}$$

then Eqs. (58) and (59)

$$r^2 \rho v c_p\, dT/dr = (d/dr)(\lambda r^2\, dT/dr) + r^2 \dot{m}_f H \tag{58}$$

$$r^2 \rho v\, dm_f/dr = (d/dr)(r^2 \rho D\, dm_A/dr) + r^2 \dot{m}_f \tag{59}$$

can be combined for the condition $\lambda/c_p = \rho D$ or $\alpha = D$ to give

$$r^2 \rho v (d/dr)(b)_{f0,\, fq,\, 0q} = (d/dr)(r^2 \rho D\, db/dr) \tag{85}$$

Equation (85), with the new definition of b, holds whether chemical reaction does or does not take place above the droplet surface. Since $\rho D = \lambda/c_p$ is considered independent of temperature, Eq. (85) may be integrated twice with the boundary conditions

$$D(db/dr)_s = v_s \quad \text{at} \quad r = r_s$$

and

$$b = b_\infty \quad \text{at} \quad r \to \infty \tag{86}$$

to obtain the solution

$$r^2 \rho v / r D \rho = \ln\{(b_\infty - b_s + 1)/(b - b_s + 1)\} \tag{87}$$

Again recall that the mass burning rate $\dot{m}_f$ is a constant so that $r^2 \rho v$ is a constant. At $r = r_s$,

$$r_s \rho_s v_s / D_s \rho_s = \ln(b_\infty - b_s + 1) = \ln(1 + B) \tag{88}$$

$$4\pi r_s^2 \rho_s v_s / 4\pi r_s D_s \rho_s = \ln(1 + B)$$

$$\dot{m}_f / 4\pi r_s = D_s \rho_s \ln(1 + B) = (\lambda/c_p) \ln(1 + B) \tag{89}$$

$$G_f = \dot{m}_f / 4\pi r_s^2 = (D_s \rho_s / r_s) \ln(1 + B)$$

As in the case of evaporation, it is important to note that since $D\rho$ is independent of pressure, the burning rate of a droplet in a quiescent atmosphere is also independent of the pressure. The transfer number B for these cases can take any of the following forms:

without combustion assumption — with combustion assumption

$$B_{f0} = \frac{(m_{f\infty} - m_{fs}) + (m_{0s} - m_{0\infty})i}{(m_f s - 1) + i m_{0s}} = \frac{i m_{0\infty} + m_{fs}}{1 - m_{fs}}$$

$$B_{fq} = \frac{H(m_{f\infty} - m_{fs}) + c_p(T_\infty - T_s)}{L_v + H(m_{fs} - 1)} = \frac{c_p(T_\infty - T_s) - m_{fs} H}{L_v + H(m_{fs} - 1)} \tag{90}$$

$$B_{0q} = \frac{Hi(m_{0\infty} - m_{0s}) + c_p(T_\infty - T_s)}{L + i m_{0s} H} = \frac{c_p(T_\infty - T_s) + i m_{0\infty} H}{L_v}$$

The combustion assumption in Eqs. (90) is that $m_{0s} = m_{f\infty} = 0$ since it is assumed that neither the fuel or oxidizer can penetrate the flame zone. This requirement is not that the flame zone must be infinitely thin, but that all the oxidizer must be consumed before it reaches the surface and that there be no fuel in the quiescent atmosphere.

In order to solve Eq. (89), it is necessary to proceed as in the preceding section and first equate $B_{f0} = B_{0q}$. This expression is solved with use of vapor pressure data for the fuel. The solution of $B_{f0} = B_{0q}$ gives T_s and m_{fs} and thus individually B_{f0} and B_{0q}. With B known, then the burning rates are obtained from Eq. (89).

For most combustion systems, B_{0q} is the most convenient form of B. $c_p(T_\infty - T_s)$ is usually very much less than $im_{0\infty} H$ and to a close approximation $B_{0q} \cong im_{0\infty} H/L_v$, and the burning rate (or as will be shown later β) is readily determined.

Values of B taken from both Spalding (1955) and Blackshear (1960) for various condensed combustible substances are given in Table 1. Examination of this table reveals that, excluding carbon, variations in B for different combustible materials are not great. Indeed, for most industrial fuels it varies only from 5.5 to 1.7. Since the transfer number always enters the burning rate expression (and the expression for β to be determined later) as a $\ln(1 + B)$ term, one may conclude that as long as the diffusion atmosphere is kept the same, there will not be a great variation in the burning rate or β of various materials. A tenfold variation in B results only in an approximately twofold variation in burning rate. A tenfold variation in the diffusivity or gas density would result in a tenfold increase in burning rate.

TABLE 1

Values of the transfer number in air

Combustible	B	Combustible	B
iso-Octane	6.41	Kerosene	~3.4
Benzene	5.97	Gas oil	~2.5
n-Heptane	5.82	Light fuel oil	~2.0
Toluene	5.69	Heavy fuel oil	1.7
Aviation gasoline	~5.5	Carbon	0.12
Automobile gasoline	~5.3		

Further, it is most interesting to note that the burning rate has been determined without determining the flame temperature or the position of the flame. In the approach attributed to Godsave (1953), it was necessary to find the flame temperature, and much of the early burning rate developments followed this procedure. In the early literature there are frequent comparisons not only of the calculated and experimental burning rates (or β), but also of the flame temperature and position. To their surprise, most experimenters found, good agreement with respect to burning rate, but poorer agreement in flame temperature and position. What they failed to realize is that, as shown by the developments here, the burning rate is independent of the flame temperature. As long as an integrated approach is used and the gradients of temperature and product concentration must be zero at the outer boundary, then it does not matter what or where the reactions take place so long as they take place within the boundaries of the integration.

It is possible to determine the flame temperature T_f and position r_f that would correspond to the Godsave-type calculations, simply by assuming the flame exists at the position where $im_0 = m_f$. Equation (87) is written in terms of b_{f0} as

$$r_s^2 \rho_s v_s / D_s \rho_s r = \ln\left(\frac{m_{f\infty} - m_{fs} - i(m_{0\infty} - m_{0s}) + (m_{fs} - 1)}{m_f - m_{fs} - i(m_0 - m_{0s}) + (m_{fs} - 1)}\right) \tag{91}$$

At the flame surface, $m_f = m_0 = 0$ and $m_{f\infty} = m_{0s} = 0$, thus Eq. (91) becomes

$$r_s^2 \rho_s v_s / D_s \rho_s r_f = \ln(1 + im_{0\infty}) \tag{92}$$

Since $\dot{m}_f = 4\pi r_s^2 \rho_s v_s$ is known, r_f can be determined. The flame temperature T_f at r_f can be obtained by writing Eq. (88) with $b = b_{0q}$. Making use of Eq. (92),

$$1 + im_{0\infty} = (Him_{0\infty} + c_p(T_\infty - T_s) + L_v)/(c_p(T_f - T_s) + L_v)$$

or

$$c_p(T_f - T_s) = (Him_{0\infty} + c_p(T_\infty - T_s) + L_v)/(1 + im_{0\infty}) - L_v \tag{93}$$

Although it is now possible to calculate the burning rate of a droplet under the quasi-steady conditions outlined and to estimate as well the flame temperature and position, the only means to estimate the burning time of an actual droplet is to calculate the evaporation coefficient for burning β. From the mass burning results obtained, β may be readily determined. For a liquid droplet, the relation

$$dm/dt = -\dot{m}_f = 4\pi\rho_1 r_s^2\, dr_s/dt \tag{94}$$

gives the mass rate in terms of the change of radius with time. ρ_1 is the density of the liquid fuel. It should be evaluated at T_s.

Many experimenters have obtained results similar to those given in Fig. 8. These results confirm that the variation of droplet diameter during burning follows the same "law" as that for evaporation

$$d^2 = d_0^2 - \beta t \tag{95}$$

Since $d = 2r_s$,

$$dr_s/dt = -\beta/8r_s$$

It is readily shown that Eqs. (89) and (94) verify that a "d^2" law should exist. Equation (94) is rewritten as

$$\dot{m} = -2\pi\rho_1 r_s\, dr^2/dt = -(\pi\rho_1 r_s/2)(d(d^2)/dt)$$

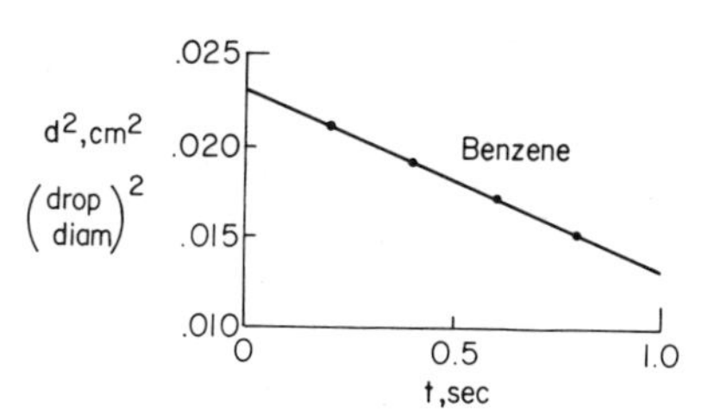

Fig. 8. Benzene droplet in quiescent air showing diameter-squared time dependency (after Godsave, 1953).

Making use of Eqs. (89),

$$\dot{m}/4\pi r_s = (\lambda/c_p)\ln(1+B) = -(\rho_l/8)(d(d^2)/dt) = -(\rho_l/8)\beta$$

shows that

$$d(d^2)/dt = \text{const}$$

and

$$\beta = (8/\rho_l)(\lambda/c_p)\ln(1+B) \tag{96}$$

which is a convenient form since λ/c_p is temperature insensitive. Sometimes β is written as

$$\beta = 8(\rho_s/\rho_l)\alpha\ln(1+B)$$

to correspond more closely to expressions given by Eqs. (81) and (88).

C. BURNING IN CONVECTIVE ATMOSPHERES

1. The Stagnant Film Case

A spherical–symmetric fuel droplet burning problem is the only quiescent case which is mathematically tractable. However the equations for mass burning may be readily solved in one-dimensional form for what may be considered the stagnant film case. If the stagnant film is of thickness δ, then the free stream conditions are thought to exist at some distance δ from the fuel surface (see Fig. 9).

Within the stagnant film, the energy equation may be written as

$$\rho v c_p\, dT/dy = \lambda(d^2T/dy^2) + \dot{q} \tag{97}$$

Defining b as before, the solution of this equation and case proceeds as follows. Analogous to Eq. (85), for the one-dimensional case

$$\rho v\, db/dy = \rho D\, d^2b/dy^2 \tag{98}$$

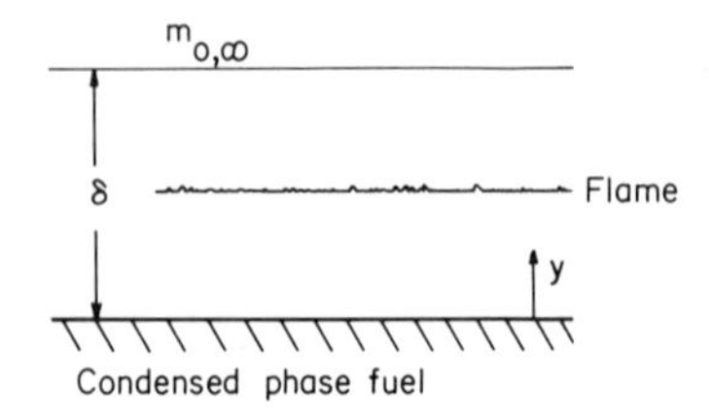

Fig. 9. Stagnant film height for condensed phase burning.

Integrating Eq. (98)

$$\rho v b = \rho D\, db/dy + \text{const} \tag{99}$$

The boundary condition is

$$\rho D(db/dy)_0 = \rho_s v_s = \rho v \tag{100}$$

Substituting this boundary condition into Eq. (99)

$$\rho v b_0 = \rho v + \text{const}, \qquad \rho v(b_0 - 1) = \text{const}$$

The integrated equation now becomes

$$\rho v(b - b_0 + 1) = \rho D\, db/dy \tag{101}$$

which upon second integration becomes

$$\rho v y = \rho D \ln(b - b_0 + 1) + \text{const} \tag{102}$$

At $y = 0$, $b = b_0$, therefore the constant equals zero so that

$$\rho v y = \rho D \ln(b - b_0 + 1) \tag{103}$$

Since δ is the stagnant film thickness,

$$\rho v \delta = \rho D \ln(b_\delta - b_0 + 1) \tag{104}$$

Thus

$$G_f = (\rho D/\delta) \ln(1 + B) \tag{105}$$

where, as before,

$$B = b_{\delta \text{ or } \infty} - b_0 \tag{106}$$

Since the Prandtl number $c_p \mu/\lambda$ is equal to 1, Eq. (105) may be written as

$$G_f = (\rho D/\delta) \ln(1 + B) = (\lambda/c_p \delta) \ln(1 + B) = (\mu/\delta) \ln(1 + B) \tag{107}$$

A burning pool of liquid or a volatile solid fuel will establish a stagnant film height due to the natural convection which ensues. From analogies to heat transfer without mass transfer, a first approximation to the liquid pool burning rate may be written as

$$G_f\, d/\mu \ln(1 + B) \sim \mathrm{Gr}^a \tag{108}$$

where a equals $\frac{1}{4}$ for laminar conditions and $\frac{1}{3}$ for turbulent conditions, d is the critical dimension of the pool, and Gr is the Grashof number.

$$\mathrm{Gr} = (gd^3\beta_1/\alpha^2)\,\Delta T \tag{109}$$

where g is the gravitational constant and β_1 the coefficient of expansion.

Whenever the free stream, whether forced or due to buoyancy effects, is transverse to the mass evolution from the regressing fuel surface, a stagnant film does not exist, and the correlation given by Eq. (108) would not be explicitly correct.

2. The Longitudinally Burning Surface

Many practical cases of burning in convective atmospheres may be approximated by consideration of a burning longitudinal surface. This problem is similar to what could be called the burning flat plate case and has application to such problems as that which occur in the hybrid rocket. It differs from the stagnant film case in that there are gradients in the x-direction as well as the y-direction. This configuration is depicted in Fig. 10.

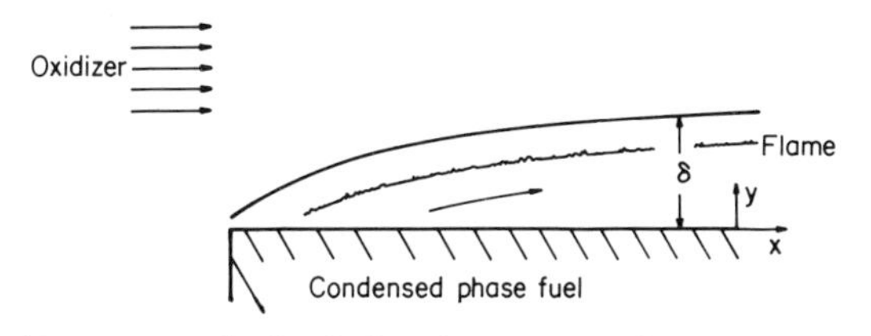

Fig. 10. Burning of a flat fuel surface in a one-dimensional flow field.

For the case of Schmidt number equal to 1, it can be shown (see Williams, 1965) that the conservation equations (in terms of Ω, see Eq. (17)) can be transposed into the same form as the momentum equation for the boundary layer. Indeed, the transformations give the same form as the incompressible boundary layer equations developed and solved by Blasius. The important difference between this problem and the Blasius (1908) problem is the boundary condition at the surface. The Blasius equation takes the form

$$f''' + ff'' = 0 \tag{110}$$

where f is a modified stream function and the primes designate differentiation with respect to a transformed coordinate. The boundary conditions at the surface for the Blasius problem are

$$\begin{aligned} \eta = 0, &\quad f(0) = 0 \quad \text{and} \quad f'(0) = 0 \\ \eta = \infty, &\quad f'(\infty) = 1 \end{aligned} \tag{111}$$

For the mass burning problem, the boundary conditions at the surface are

$$\begin{aligned} \eta = 0, &\quad f(0) = 0 \quad \text{and} \quad Bf''(0) = -f(0) \\ \eta = \infty, &\quad f'(\infty) = 1 \end{aligned} \tag{112}$$

where η is the transformed distance normal to the plate. The second of these conditions contains the transfer number B and is of the same form as the boundary condition in the stagnant film case (Eq. (100)).

Emmons (1956) solved this burning problem and his results can be shown to be of the form

$$G_f = (\lambda/c_p)(\mathrm{Re}_x^{1/2}/x\sqrt{2})[-f(0)] \tag{113}$$

where Re_x is the Reynold's number based on the distance x from the leading edge of the flat plate. For Prandtl number equal to 1, Eq. (113) can be written in the form

$$G_f x/\mu = \mathrm{Re}_x^{1/2}[-f(0)]/\sqrt{2} \tag{114}$$

It is particularly important to note that $[-f(0)]$ is a function of the transfer number B. This dependency as determined by Emmons is shown in Fig. 11.

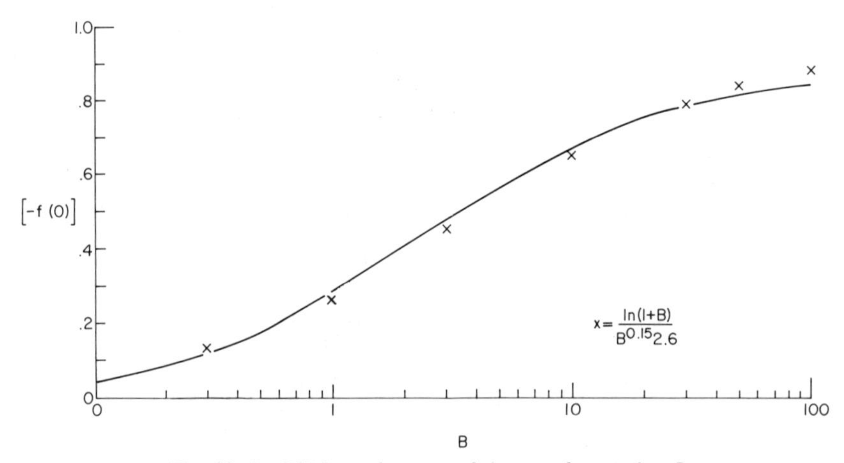

Fig. 11. $[-f'(0)]$ as a function of the transfer number B.

An interesting approximation to this result (Eq. (114)) can be made from the stagnant film result of the last section, i.e.,

$$G_f = (D\rho/\delta)\ln(1+B) = (\lambda/c_p\delta)\ln(1+B) = (\mu/\delta)\ln(1+B) \quad (107)$$

If δ were assumed to be the Blasius boundary layer thickness δ_x, then

$$\delta_x = 5.2x\,\mathrm{Re}_x^{-1/2} \quad (115)$$

Substituting Eq. (115) into Eq. (107),

$$G_f x/\mu = (\mathrm{Re}_x^{1/2}/5.2)\ln(1+B) \quad (116)$$

The values predicted by Eq. (116) are somewhat high compared to those predicted by Eq. (114). If Eq. (116) is divided by $B^{0.15}/2$ to give

$$G_f x/\mu = (\mathrm{Re}_x^{1/2}/2.6)(\ln(1+B)/B^{0.15}) \quad (117)$$

then the agreement is very good over a wide range of B values. To show the extent of the agreement, the function

$$\ln(1+B)/B^{0.15}2.6 \quad (118)$$

is plotted on Fig. 11 as well.

Obviously these results do not hold at very low Reynold's number. As Re approaches zero, the boundary layer thickness approaches infinity. However, the burning rate is bounded by the quiescent results.

3. The Flowing Droplet Case

When droplets are not at rest relative to the oxidizing atmosphere, the quiescent results no longer hold and forced convection must again be considered. No one has solved this complex case. As discussed in Section 4.E, flow around a sphere can be complex, and at relatively high Re (>20), there is a boundary layer flow around the front of the sphere and a wake or eddy region behind it.

For this burning droplet case, an overall heat transfer relationship could be written to define the boundary condition given by Eq. (75)

$$h(\Delta T) = G_f L \quad (119)$$

The thermal driving force is represented by a temperature gradient ΔT which is the ambient temperature T_∞ plus the rise in this temperature due to the energy release minus the temperature at the surface T_s or

$$\Delta T = T_\infty + (im_{0\infty}H/c_p) - T_s = [im_{0\infty}H + c_p(T_\infty - T_s)]/c_p \quad (120)$$

Substituting Eq. (120) and Eq. (89) for G_f into Eq. (119),

$$h[im_{0\infty}H + c_p(T_\infty - T_s)]/c_p = [(D\rho/r)\ln(1+B)]L = [(\lambda/c_p r)\ln(1+B)]L \tag{121}$$

where r is now the radius of the droplet. Cross-multiplying, Eq. (121) becomes

$$hr/\lambda = \ln(1+B)/B = \mathrm{Nu} \tag{122}$$

since

$$B = [im_{0\infty}H + c_p(T_\infty - T_s)]/L$$

Since Eq. (89) was used for G_f, this Nusselt number (Eq. (122)) is for the quiescent case ($\mathrm{Re} \to 0$). For small B, $\ln(1+B) \approx B$ and the $\mathrm{Nu} = 1$, the classical result for heat transfer without mass addition.

The term $[\ln(1+B)]/B$ has been used as an empirical correction for higher Reynold's number problems and a classical expression for Nu with mass transfer is

$$\mathrm{Nu}_r = (\ln(1+B)/B)[1 + 0.39\,\mathrm{Pr}^{1/3}\,\mathrm{Re}_r^{1/2}] \tag{123}$$

As $\mathrm{Re} \to 0$, Eq. (123) approaches Eq. (122). For the case $\mathrm{Pr} = 1$ and $\mathrm{Re} \gg 1$, Eq. (123) becomes

$$\mathrm{Nu}_r = (.39)[\ln(1+B)/B]\mathrm{Re}_r^{1/2} \tag{124}$$

The flat plate result of the preceding section could have been written in terms of a Nusselt number as well. In that case

$$\mathrm{Nu}_x = [-f(0)]\mathrm{Re}_x^{1/2}/\sqrt{2}B \tag{125}$$

Thus the burning rate expressions related to Eqs. (124) and (125) are, respectively,

$$G_f r/\mu = 0.39\,\mathrm{Re}_r^{1/2}\ln(1+B) \tag{126}$$

$$= \mathrm{Re}_r^{1/2}[-f(0)]/\sqrt{2} \tag{127}$$

Since in convective flow a wake exists behind the droplet and droplet heat transfer in the wake may be minimal, these equations are not likely to predict quantitative results. It is again interesting to note that if the right-hand side of Eq. (126) is divided by $B^{0.15}$, then the expression given by Eqs. (126) and (127) follow identical trends and thus data can be correlated as

$$(G_f r/\mu)\{B^{0.15}/\ln(1+B)\} \text{ vs } \mathrm{Re}^{1/2} \tag{128}$$

If turbulent boundary layer conditions are achieved under certain conditions, then the same type of expression should hold and Re should be raised to the 0.8 power.

If, indeed, Eqs. (126) and (127) adequately predict the burning rate of a droplet in laminar convective flow, then for a given relative velocity between the gas and the droplet, the droplet will follow a "$d^{3/2}$" burning rate law. β in this case will be a function of the relative velocity as well as B and other physical parameters of the system. This result should be compared to the "d^2" law (Eq. (95)) for droplet burning in quiescent atmospheres. In turbulent flow, droplets will appear to follow a burning rate law in which the power of the diameter is close to one.

4. Burning Rates of Plastics; The Small B Assumption and Radiation Effects

There is great interest in the fire safety of polymeric (plastic) materials and in determining the mass burning rate of plastics. For plastics whose burning rate measurements are made so that there is no melting, or for nonmelting plastics, the developments just obtained should hold well. For the burning of some plastics in air or at low oxygen concentrations, the transfer number may be considered small compared to 1. For this condition, which of course would hold for any system in which $B \ll 1$,

$$\ln(1 + B) \cong B \tag{129}$$

and the mass burning rate expression may be written for nontransverse movement of the air case

$$G_{\mathrm{f}} \cong (\lambda/c_{\mathrm{p}}\delta)B \tag{130}$$

Recall that for these considerations the most convenient expression for B is

$$B = im_{0\infty} H + c_{\mathrm{p}}(T_{\infty} - T_{\mathrm{s}})/L_{\mathrm{v}} \tag{131}$$

In most cases

$$im_{0\infty} H \gg c_{\mathrm{p}}(T_{\infty} - T_{\mathrm{s}}) \tag{132}$$

so

$$B \cong im_{0\infty} H/L_{\mathrm{v}} \tag{133}$$

and

$$G_{\mathrm{f}} \cong (\lambda/c_{\mathrm{p}}\delta)(im_{0\infty} H/L_{\mathrm{v}}) \tag{134}$$

Equation (134) shows the reason that for burning rate experiments in which the dynamics of the air are constant or well controlled (i.e., δ is known or constant), then good straightline correlations are obtained when G_f is plotted as a function of $m_{0\infty}$. One should realize that

$$G_f \sim m_{0\infty} \tag{135}$$

holds only for small B.

The consequence of this small B assumption may not at first be apparent. A physical interpretation can be obtained from again writing the mass burning rate expression for the two assumptions made; i.e., $B \ll 1$ and $B \cong (im_{0\infty} HL_v)$:

$$G_f \cong (\lambda/c_p \delta)(im_{0\infty} H/L_v) \tag{136}$$

and realizing that as an approximation

$$H \cong c_p(T_f - T_s) \tag{137}$$

where T_f is the diffusion flame temperature. Thus the burning rate expression becomes

$$G_f \cong (\lambda/c_p \delta)(c_p(T_f - T_s)/L_v) \tag{138}$$

Cross-multiplying, one obtains

$$G_f L_v \cong \lambda(T_f - T_s)/\delta \tag{139}$$

which says that the energy required to gasify the surface at a given rate per unit area is supplied by the heat flux established by the flame. Equation (139) is simply another form of Eq. (119). Thus, the significance of the small B assumption is that the gasification from the surface is so small that it does not alter the gaseous temperature gradient determining the heat flux to the surface. This result is different from that obtained earlier and which stated that the stagnant film thickness was not affected by the surface gasification rate. The small B assumption goes one step further and reveals that under this condition the temperature profile in the boundary layer is not affected by the small amount of gasification.

If in the mass burning process there is flame radiation, or any other imposed radiation, as is frequently used in plastic flammability tests, then a convenient expression for the mass burning rate can be obtained provided it is assumed that only the gasifying surface and not any of the gases between the radiation source and the surface absorbs radiation. In this case Fineman (1962) showed that the stagnant film expression for the burning rate can be approximated by

$$G_f = (\lambda/c_p \delta) \ln[1 + (B/1 - E)] \qquad \text{where} \quad E \equiv Q_R/G_f L_v$$

and Q_R is the radiative heat transfer flux.

This simple form for the burning rate expression arises because the equations are developed for the conditions in the gas phase and the mass burning rate arises explicitly in the boundary condition to the problem. Since the assumption is made that no radiation is absorbed by the gases, the radiation term appears only in the boundary condition to the problem.

Notice that as the radiant flux Q_R increases, E and the term $[B/(1-E)]$ increase. When $E = 1$, the problem disintegrates because the equation was developed in the framework of a diffusion analysis. $E = 1$ means that the solid is gasified by the radiant flux alone.

REFERENCES

Blackshear, P. L. Jr. (1960). "An Introduction to Combustion," Chapter IV. Dept. of Mech. Eng., Univ. of Minnesota, Minneapolis, Minnesota.
Blasius, H. (1908). *Z. Math. Phys.* **56**, 1.
Burke, S. P., and Schumann, T. E. W. (1928). *Ind. Eng. Chem.* **20**, 998.
Emmons, H. W. (1956). *Z. Angew. Math. Mech.* **36**, 60.
Fineman, S. (1962). M.S.E. Thesis, Dept. Aero. Eng., Princeton Univ., Princeton, New Jersey.
Godsave, G. A. E. (1953). *Int. Symp. Combust., 4th* p. 818. Williams and Wilkens, Baltimore, Maryland.
Hottell, H. C., and Hawthorne, W. R. (1949). *Int. Symp. Combust., 3rd* p. 255. Williams and Wilkens, Baltimore, Maryland.
Khudyakov, L. (1955). *Chem. Abstr.* **46**, 10844e.
Lewis, B., and von Elbe, G. (1961). "Combustion, Flames and Explosions of Gases," 2nd ed., Chapter VII. Academic Press, New York.
Spalding, D. B. (1953). *Int. Symp. Combust., 4th* p. 847. Williams and Wilkens, Baltimore, Maryland.
Spalding, D. B. (1955). "Some Fundamentals of Combustion," Chapter 4. Butterworths, London.
Williams, F. A. (1965). "Combustion Theory," Chapters 1, 3, and 12. Addison-Wesley, Reading, Massachusetts.

Chapter Seven

Ignition

A. CONCEPTS

If only gas phase combustion were to be treated in this work, then the subject of ignition should characteristically be treated prior to the discussion of gaseous explosions (Chapter 3). However, ignition of condensed phases is equally significant and should be discussed. The problem of the storage of wet coal or large concentrations of solid materials which can undergo slow exothermic decomposition is also one of ignition. Indeed the concept of spontaneous combustion is an element of the theory of thermal ignition.

Ignition is considered the process of producing an explosion, a deflagration, detonation, or any diffusion flame phenomena. In a pure adiabatic system, if one assumes that an Arrhenius expression defines the kinetics, then the system is reacting at every temperature no matter how slowly. In this case it is not possible to define uniquely the ignition temperature. It has been the practice, however, to define an ignition time for a system at a given temperature in order to establish a procedure consistent with some theories of ignition. This time, then, in order to be a more uniquely defined parameter, is usually meant to specify that interval for a given initial fraction of reactants to be consumed.

Many early experiments report the ignition temperature of various liquid fuels in air by observing the air temperature at which a droplet would break into a flame. The importance of the time element was not recognized. In these experiments, however, there was an inherent time element. This time is that mentioned above and is the time it takes for the reaction of the fuel vapor of the droplet and the air to reach a degree of reaction and thus temperature, so that the products are luminous. Thus while it would be correct to define an ignition temperature as the initial air temperature which causes the reaction to come to a given fractional completion in a given time, one cannot permit the individual gas phase kinetics of the fuel–oxygen system to govern the time. The droplet ignition experiments give useful results because, fortunately, they are performed mostly on hydrocarbons whose reaction kinetics do not vary too severely.

When there are heat losses, the nonadiabatic consideration, one can give a more reasonable definition of the ignition temperature as the lowest temperature at which the rate of heat loss from the system is overbalanced by the rate of heat generation by chemical reaction. Here, the time is generally not important since very long times can prevail before a rapid rise in temperature is observed (i.e., a explosion or flame is generated). The important critical parameter here is one related to heat loss, and, as will be shown later, is a critical dimension. This problem is the critical mass or spontaneous ignition problem. An interesting example is the explosion of stored ammonium nitrate (see Hainer, 1955).

In real cases ignition is truly a chemical phenomenon which is related to the physical phenomenon of heat transfer. Any experiment which does not precisely define the heat transfer (loss) conditions or introduces other physical factors such as mixing, turbulence, etc., is not adequate for ignition determinations.

B. THE THEORY OF THERMAL IGNITION

The concepts of adiabatic and nonadiabatic ignition can best be approached by dealing with what occurs to a combustible mixture when it is contained in a vessel whose walls are at a temperature T_0. Essentially this description fits the experiment depicted in Fig. 3.1 and underlies the concepts of the thermal explosion mentioned frequently in Chapter 3.

From a strictly thermal point, in contrast to the chain-branching kinetics point of view discussed extensively in Chapter 3, and according to the conditions of mixture ratio, size, and temperature T_0, the combustible mixture can undergo a rapid growth in temperature and reach its adiabatic flame temperature. Under other conditions, only an insignificant rise in temperature to a stationary level not too different from T_0 will be observed. This stationary temperature rise will remain constant until a considerable part of the mixture has reacted. The conditions under which the stationary temperature range goes to the explosive range are termed the critical conditions of ignition. The very existence of a sharp transition is, of course, contingent on certain conditions which must be formulated.

In order to define the conditions so that the clearest understanding is obtained, it is proper to consider that there are no convective motions within the vessel. This condition is certainly valid for solids, but is an assumption for gaseous mixtures or liquids. Following the methods of Frank-Kamenetskii (1955), the equation for pure heat conduction within the vessel becomes the starting point. However, since the gaseous mixture, liquid or solid, can undergo exothermic transformations, a chemical reaction term must be included. This term specifies a continuously distributed source of heat throughout the vessel. The heat conduction equation for the vessel is then

$$c_p \rho \, dT/dt = \operatorname{div}(\lambda \operatorname{grad} T) + q' \tag{1}$$

in which the nomenclature is apparent perhaps except for q' which represents the heat release density.

The solution of this equation would give the temperature distribution as a function of the spatial distance and the time. At the ignition condition, the character of this temperature distribution changes sharply. There should be an abrupt transition from a small stationary rise to a large and rapid rise. The difficulties in solving this partial differential equation are great and attempts have been made. But much insight into overall practical ignition phenomena can be gained by considering the two approximate methods as Frank-Kamenetskii has done. The two approximate methods are referred to the stationary and nonstationary solutions. In the stationary theory, only the temperature distribution throughout the vessel is considered and the time variation is ignored. In the nonstationary theory, the spatial temperature variation is not taken into account, a mean temperature value throughout the vessel is used, and the variation of this mean temperature with time is examined.

1. The Stationary Solution—The Critical Mass and Spontaneous Ignition Problems

The stationary theory deals with time-independent equations of heat conduction with distributed sources of heat. Its solution gives the stationary temperature distribution in the reacting mixture. The initial conditions under which such a stationary distribution becomes impossible are the critical conditions for ignition.

Under this steady assumption, Eq. (1) becomes

$$\operatorname{div}(\lambda \operatorname{grad} T) = -q' \tag{2}$$

and if the temperature dependency of the thermal conductivity is neglected,

$$\lambda \nabla^2 T = -q' \tag{3}$$

It is important to consider the definition of q'. It is the amount of heat evolved by chemical reaction in a unit volume and in unit time. It is the product of the terms involving the energy content of the fuel and its rate of reaction. The rate of the reaction can be written as $Ze^{-E/RT}$. Recall Z is different from the normal Arrhenius preexponential term in that it contains the concentration terms and thus can be dependent on the mixture composition and the pressure. Thus

$$q' = QZe^{-E/RT} \tag{4}$$

where Q is the volumetric energy release of the combustible mixture. It follows then that

$$\nabla^2 T = -(Q/\lambda)Ze^{-E/RT} \tag{5}$$

and the problem resolves itself in solving this equation under the boundary condition that $T = T_0$ at the wall of the vessel.

Since the stationary temperature distribution below the explosion limit is sought and thus the temperature rise throughout the vessel is to be small, it is best to introduce a new variable

$$v = T - T_0$$

in which $v \ll T_0$. Under this condition, it is possible to describe the cumbersome exponential term as

$$\exp(-E/RT) = \exp[-E/R(T_0 + v)] = \exp\left\{-\frac{E}{RT_0}\left[\frac{1}{1 + (v/T_0)}\right]\right\}$$

If the term in brackets is expanded and the higher-order terms are eliminated, the expression simplifies to

$$\exp[-E/RT] \cong \exp\left[-\frac{E}{RT_0}\left(1 - \frac{v}{T_0}\right)\right] = \exp\left[-\frac{E}{RT_0}\right]\exp\left[\frac{E}{RT_0^2}v\right]$$

and Eq. (5) becomes

$$\nabla^2 v = -\frac{Q}{\lambda}Z\exp\left[-\frac{E}{RT_0}\right]\exp\left[\frac{E}{RT_0^2}v\right] \tag{6}$$

In order to solve Eq. (6), new variables are defined

$$\theta = (E/RT_0^2)v, \qquad \eta_x = x/r$$

where r is the radius of the vessel and x the distance from the center. Equation (6) then becomes

$$\nabla_\eta \theta = -(Q/\lambda)(E/RT_0{}^2)r^2 Ze^{-(E/RT_0)}e^\theta \tag{7}$$

and the boundary condition is $\eta = 1$, $\theta = 0$.

Both Eq. (7) and the boundary condition contain only one nondimensional parameter δ:

$$\delta = (Q/\lambda)(E/RT_0{}^2)r^2 Ze^{-E/RT_0} \tag{8}$$

The solution of Eq. (7) which represents the stationary temperature distribution should be of the form $\theta = f(\eta, \delta)$ with one parameter, i.e., δ. The condition under which such a stationary temperature distribution ceases to be possible; i.e., the critical condition of ignition, is of the form $\delta = \text{const} = \delta_{\text{crit}}$. The critical value depends upon T_0, the geometry (if the vessel is nonspherical), and the pressure through Z. Numerical integration of Eq. (7) for various δ's determines the critical δ. For a spherical vessel $\delta = 3.32$; for an infinite cylindrical vessel, $\delta = 2.00$; and for infinite parallel plates, $\delta = 0.88$.

As in the discussion of flame propagation, the pressure dependency in Z is $Z \sim p^{n-1}$, where n is the order of the reaction. Equation (8) expressed in terms of δ_{crit} permits the relationship between the critical parameters to be determined. Taking logarithms,

$$\ln rp^{n-1} \sim (+E/RT_0)$$

If the reacting medium is a solid or liquid undergoing exothermic decomposition, then the pressure term is omitted and

$$\ln r \sim (+E/RT_0)$$

These results define the conditions for the critical size of storage for compounds such as ammonium nitrate as a function of the ambient temperature T_0. Similarly it represents the variation of a size of combustible material that will spontaneously ignite as a function of the ambient temperature T_0. The higher the ambient temperature, the smaller the critical mass has to be to prevent disaster.

2. The Nonstationary Solution

The nonstationary theory deals with the thermal balance of the whole reaction vessel and assumes the temperature to be the same at all points. This assumption is of course incorrect in the conduction range where the temperature gradient is by no means localized at the wall. It is, however, equivalent to a replacement of the mean values of all temperature-dependent magnitudes by their values at a mean temperature, and involves relatively minor error.

If the volume of the vessel is designated by V, its surface area by A, and if a heat transfer coefficient h is defined, then the amount of heat evolved over the whole volume per unit time by the chemical reaction is

$$VQZe^{-E/RT} \tag{9}$$

and the amount of heat carried away from the wall is

$$hA(T - T_0) \tag{10}$$

Thus the problem is now essentially nonadiabatic. The difference between the two heat terms is the heat which causes the temperature within the vessel to rise a given amount per unit time, i.e.,

$$c_p \rho V \, dT/dt \tag{11}$$

These terms can be expressed as an equality,

$$c_p \rho V \, dT/dt = VQZe^{-E/RT} - hA(T - T_0) \tag{12}$$

or

$$dT/dt = (Q/c_p \rho)Ze^{-E/RT} - (hA/c_p \rho V)(T - T_0) \tag{13}$$

Equations (12) and (13) are forms of Eq. (1) with a heat loss term. Nondimensionalizing the temperature and linearizing in the exponent in the same manner as in the previous section, one obtains

$$d\theta/dt = (Q/c_p \rho)(E/RT_0{}^2)Ze^{-E/RT_0}e^{\theta} - (hA/c_p \rho V)\theta \tag{14}$$

with the boundary condition $\theta = 0$ at $t = 0$.

The equation is not in dimensionless form. Each term has the dimension of reciprocal time. In order to make the equation completely dimensionless, it is necessary to introduce a time parameter. Equation (14) contains two such time parameters

$$\tau_1 = [(Q/c_p \rho)(E/RT_0{}^2)Ze^{-E/RT_0}]^{-1}, \qquad \tau_2 = (hA/c_p \rho V)^{-1}$$

Consequently, the solution of Eq. (14) should be in the form

$$\theta = f(t/\tau_{1,2}, \tau_2/\tau_1)$$

Thus the dependence of dimensionless temperature θ on dimensionless time $t/\tau_{1,2}$ contains one dimensionless parameter τ_2/τ_1. Consequently, a sharp rise in temperature can occur for a critical value of τ_2/τ_1.

It is best to examine Eq. (14) written in terms of these parameters,

$$d\theta/dt = (e^{\theta}/\tau_1) - (\theta/\tau_2) \tag{15}$$

In the ignition range the rate of energy release is much greater than the rate of heat loss; that is, the first term on the right-hand side of Eq. (15) is much greater than the second. Under these conditions it is possible to disregard the removal of heat and view the thermal explosion as adiabatic.

Then for an adiabatic thermal explosion, the time dependence of the temperature should be of the form

$$\theta = f(t/\tau_1) \tag{16}$$

Under these conditions the time within which a given value of θ is attained is proportional to the magnitude τ_1. Consequently, the induction period in the instance of adiabatic explosion is proportional to τ_1. The proportionality has been shown to be unity. Conceptually this induction period can be related to the time period for the ignition of droplets for different air (or ambient) temperatures. This τ can be the adiabatic induction time and is simply

$$\tau = \frac{c_p \rho}{Q} \frac{R T_0^2}{E} \frac{1}{Z} e^{(+E/RT_0)} \tag{17}$$

Again the expression can be related to the critical conditions of time, pressure, and ambient temperature T_0 by taking logarithms,

$$\ln(\tau P^{n-1}) \sim (+E/RT_0) \tag{18}$$

Since in dealing with hydrocarbon combustion, the overall order is approximately 2, then

$$\ln(\tau P) \sim (E/RT_0) \tag{19}$$

which is essentially the condition used in Section 4.E. This result gives the intuitively expected answer that the higher the ambient temperature the shorter the ignition time.

The condition under which the last term in Eq. (15) may be omitted can be established in a simplified manner. The heat generation term and the heat loss term can be written in the following concise manner:

$$q_g = Ce^{-E/RT} \tag{20}$$

$$q_l = B(T - T_0) \tag{21}$$

where B and C are combinations of constant terms. Plots of the heat generation term q_g and the heat loss term q_l for given initial temperatures T_0 are shown in Fig. 1. For curve A (initial temperature T_0'), there are two intersections with the generation term. The lower point 1 represents a stable condition; i.e., the medium will self-heat to the temperature represented by point 1. It cannot rise further, because the heat loss is

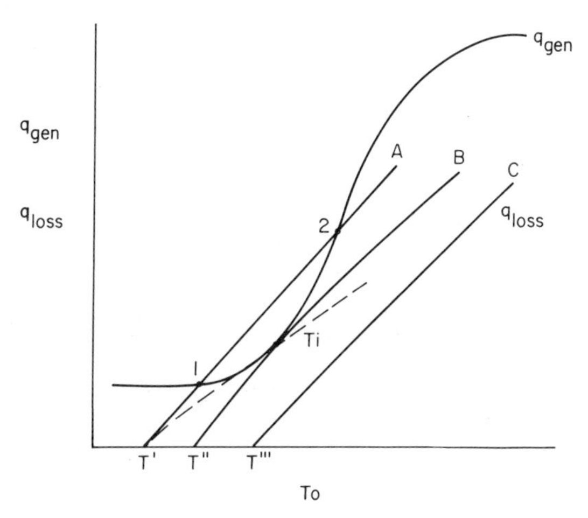

Fig. 1. Heat generation and heat loss of a reacting mixture in a vessel in a thermal bath of temperature T_0.

greater than the heat generation and any temperature perturbations above point 1 fall back to this point. One would have to heat the mixture by some other source to point 2 or above for the temperature to run away; i.e., for the mixture to explode. For curve C every point is below the generation curve and thus the reacting mixture will run away to explosion. The critical temperature for explosion is T_0'', for curve B is just tangent to the generation curve. This condition will self heat to T_i and then run away. T_i is, of course, an ignition temperature. T_0'' is then the temperature under which the self-heating explosion condition would exist and is the self-heating explosive condition discussed in Chapter 3.

The slopes of curves A, B, and C are constant and are given by B in Eq. (21). The effect of the parameters in B can easily be recognized. It would be possible, for instance, to draw a series of curves emanating from T_0' and with different slopes. One of these curves would be tangent to the generation curve and would represent an area-to-volume ratio or a vessel radius that could cause explosion when the initial temperature was T_0' and not T_0''. Thus one must conclude that the self-heating explosion limit discussed in Chapter 3 could be somewhat affected by vessel size, or for that matter by the pressure through the density.

REFERENCES

Frank-Kamenetskii, D. A. (1955). "Diffusion and Heat Exchange in Chemical Kinetics," Princeton Univ. Press, Princeton, New Jersey.

Hainer, R. M. (1955). *Int. Symp. Combust., 5th* p. 224. Van Nostrand-Reinhold, Princeton, New Jersey.

Chapter Eight

Environmental Combustion Considerations

In the mid-1940s symptoms now recognized as being due to photochemical air pollution were first encountered in Los Angeles. Several researchers identified that the conditions in the Los Angeles area were a new kind of smog caused by the action of sunlight on the oxides of nitrogen and subsequent reactions with hydrocarbons. This smog was to be contrasted with the conditions in London in the early part of this century and the polluted air disaster that struck Donora, Pennsylvania. In these latter cases, particulates and sulfur oxides from the burning coal created the unhealthy conditions. In Los Angeles the primary source of nitrogen oxides and hydrocarbons was readily identified to be the automobile. The burgeoning population and industrial growth in this and other areas led to controls not only on automobiles, but other mobile and stationary sources.

Atmosphere pollution symptoms became a concern all over the world and a much greater sensitivity to environmental conditions arose when supersonic transport developments began as part of the air transport era. Questions arose as to how the water vapor ejected by the power plants of these planes would affect the stratosphere in which they were to fly. This suggestion led to the consideration of the effects of injecting large amounts of any species on the ozone balance in the stratosphere. It is interesting that the major species now thought to affect the ozone balances are again the oxides of nitrogen.

In the case of automobiles and stationary and jet power plants, it may be possible to reduce the extent of the emissions, particularly the nitrogen oxides and hydrocarbons, so that there is no severe damage to the troposphere and stratosphere. Indeed, the fluorocarbons used in aerosol spray cans may have a more devastating effect on the stratosphere than any of the other chemical species noticed.

Other factors have compounded the environmental emissions problem. A shortage of sulfur-free fossil fuels has arisen. The so-called energy crisis demands the development of fossil fuel resources from coal, shale, and secondary and tertiary oil schemes. Fuels from these sources are known to contain fuel-bound nitrogen and sulfur. Indeed the key in the more rapid development of coal usage may be the sulfur problem. Further, the fraction of aromatics in liquid fuels derived from these new sources or synthetically developed is found to be large and, in general, such fuels have serious sooting characteristics.

This chapter seeks not only to provide better understanding of the oxidation processes of sulfur and nitrogen and the processes leading to particulate (soot) formation, but also to make use of many of the chemical concepts developed so that it is possible to analyze how these pollutants affect the atmosphere.

A. THE NATURE OF PHOTOCHEMICAL SMOG

Photochemical air pollution consists of a complex mixture of gaseous pollutants and aerosols, some of which are photochemically produced. Among the gaseous compounds are the oxidizing species

O_3	NO_2	$R{-}C({=}O){-}OONO_2$
ozone	nitrogen dioxide	peroxyacyl nitrate

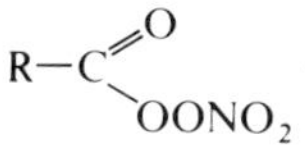

The member of this last series most commonly found in the atmosphere is

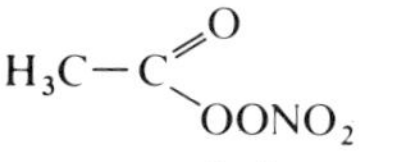

$$H_3C{-}C({=}O){-}OONO_2 \qquad \text{PAN}$$

peroxyacetyl nitrate

The three compounds, ozone, NO_2, and PAN are often grouped together and called photochemical oxidant.

In photochemical smog, one has mixtures of particulate matter and noxious gases, just as those which occur in the typical London-type smog. The London smog is a mixture of particulates and oxides of sulfur, chiefly sulfur dioxide. But the overall system is chemically reducing in nature.

This difference in redox chemistry between photochemical oxidant and SO_x-particulate smog is quite important from several aspects. In particular the problem of quantitatively detecting oxidant in the presence of sulfur dioxide is to be noted. The SO_x being a reducing agent tends to reduce the oxidizing effects of ozone and results in low quantities of the oxidant.

In dealing with the heterogeneous gas–liquid–solid mixture characterized as photochemical smog, it is important to realize from a chemical, as well as biological point of view, that synergistic effects may occur.

1. Primary and Secondary Pollutants

Primary pollutants are those emitted directly to the atmosphere and secondary pollutants are formed by chemical or photochemical reactions of primary pollutants after they have been admitted to the atmosphere and exposed to sunlight.

Unburned hydrocarbons, NO, particulates, and the oxides of sulfur are examples of primary pollutants. The particulates may be lead oxide from the oxidation of tetraethyl lead in automobiles, fly ash, and various types of carbon formations.

PAN and ozone are examples of secondary pollutants.

Some pollutants fall in both categories. NO_2, which is emitted directly from auto exhaust, is formed also in the atmosphere photochemically from NO. Aldehydes, which are released in auto exhausts, are also formed in the photochemical oxidation of hydrocarbons. CO, which arises primarily from autos and stationary sources, is again a product of atmospheric hydrocarbon oxidation.

2. The Effect of NO_x

It has been well established that if a laboratory chamber containing NO, a trace of NO_2, and air is irradiated with ultraviolet light, the following reactions occur:

$$NO_2 + h\nu(3000\ \text{Å} \leq \lambda \leq 4200\ \text{Å}) \longrightarrow NO + O(^3P) \qquad (1)$$

$$O + O_2 + M \longrightarrow O_3 + M \qquad (2)$$

$$O_3 + NO \longrightarrow O_2 + NO_2 \qquad (3)$$

The net effect of irradiation on this inorganic system is to establish the dynamic equilibrium

$$NO_2 + O_2 \underset{}{\overset{h\nu}{\rightleftharpoons}} NO + O_3 \tag{4}$$

However, if a hydrocarbon, particularly an olefin or an alkylated benzene is added to the chamber, the equilibrium represented by Eq. (4) is unbalanced and the following events take place:

a. The hydrocarbons are oxidized and disappear.
b. Reaction products such as aldehydes, nitrates, PAN, etc., are formed.
c. NO is converted to NO_2.
d. When all the NO is consumed, ozone begins to appear. On the other hand, PAN and other aldehydes are formed from the beginning.

Basic rate information permits one to examine these phenomena in detail. Leighton (1961) in his excellent book "Photochemistry of Air Pollution" gives numerous tables of rates and products of photochemical nitrogen oxide–hydrocarbon reactions in air. The data in these tables show low rates of photochemical consumption of the saturated hydrocarbons, as compared to the unsaturates, and the absence of aldehydes in the products of the saturated hydrocarbon reactions. These data conform to the relatively low rate of reaction of the saturated hydrocarbons with oxygen atoms and their inertness with respect to ozone.

Among the major products in the olefin reactions are aldehydes and ketones. Such results correspond to the splitting of the double bond and the addition of an oxygen atom to one end of the olefin.

Irradiation of mixtures of an olefin with nitric oxide and nitrogen dioxide in air shows that the nitrogen dioxide rises in concentration before it is eventually consumed by reaction. Since it is the photodissociation of the nitrogen dioxide that initiates the reaction, it would appear that a negative quantum yield results. More likely, the nitrogen dioxide is being formed by secondary reactions more rapidly than it is being photodissociated.

The important point to realize is that this negative quantum yield is only recognized when an olefin (hydrocarbon) is present. Thus adding the overall step

$$\left.\begin{matrix} O \\ O_3 \end{matrix}\right\} + \text{olefin} \longrightarrow \text{products} \tag{5}$$

to reactions (1)–(3) would not be an adequate representation of the atmosphere photochemical reactions. However, if it is assumed that O_3 attains a steady state concentration in the atmosphere, then a steady state analysis (see Section 2.B) with respect to O_3 can be performed. Further, if it is assumed that O_3 is largely destroyed by reaction (3), then a very useful approximate relationship is attained.

$$O_3 = -(j_1/k_3)(NO_2)/(NO)$$

where j is the rate constant for the photochemical reaction. Thus the O_3 steady state concentration in a polluted atmosphere is seen to increase with decreasing concentration of nitric oxide and vice versa. The ratio of j_1/k_3 approximately equals 1.2 ppm for the Los Angeles noonday condition (Leighton, 1961).

Reactions such as

$$O + NO_2 \longrightarrow NO + O_2$$

$$O + NO_2 + M \longrightarrow NO_3 + M$$

$$NO_3 + NO \longrightarrow 2NO_2$$

$$O + NO + M \longrightarrow NO_2 + M$$

$$2NO + O_2 \longrightarrow 2NO_2$$

$$NO_3 + NO_2 \longrightarrow N_2O_5$$

$$N_2O_5 \longrightarrow NO_3 + NO_2$$

do not play a part. They are generally too slow to be important.

Further, it has been noted that when the rate of the oxygen atom–olefin reaction and the rate of the ozone–olefin reaction are totaled, they do not give the complete hydrocarbon consumption. This anomaly is also an indication of an additional process.

An induction period with respect to olefin consumption is also noted in the photochemical laboratory experiments and indicates the buildup of an intermediate. When illumination is terminated in these experiments, the excess rate over the total of the O and O_3 reactions disappears. These and other results indicate that the intermediate formed is photolyzed and contributes to the concentration of the major species of concern.

Possible intermediates which fulfil the requirements of the laboratory experiments are alkyl and acyl nitrites and pernitrites. The second photolysis effect eliminates the possibility of aldehydes being the intermediate.

Various mechanisms have been proposed to explain the laboratory results discussed above. The following low (atmospheric) temperature sequence based on isobutene as the initial fuel has been proposed by Leighton (1961) and appears to account for most of what has been observed:

$$O + C_4H_8 \longrightarrow CH_3 + C_3H_5O \tag{6}$$

$$CH_3 + O_2 \longrightarrow CH_3OO \tag{7}$$

$$CH_3OO + O_2 \longrightarrow CH_3O + O_3 \tag{8}$$

$$O_3 + NO \longrightarrow NO_2 + O_2 \quad (9)$$
$$CH_3O + NO \longrightarrow CH_3ONO \quad (10)$$
$$CH_3ONO + h\nu \longrightarrow CH_3O^* + NO \quad (11)$$
$$CH_3O^* + O_2 \longrightarrow H_2CO + HOO \quad (12)$$
$$HOO + C_4H_8 \longrightarrow H_2CO + (CH_3)_2CO + H \quad (13)$$
$$M + H + O_2 \longrightarrow HOO + M \quad (14)$$
$$HOO + NO \longrightarrow OH + NO_2 \quad (15)$$
$$OH + C_4H_8 \longrightarrow (CH_3)_2CO + CH_3 \quad (16)$$
$$CH_3 + O_2 \longrightarrow CH_3OO \quad \text{as above} \quad (17)$$
$$2HOO \longrightarrow H_2O_2 + O_2 \quad (18)$$
$$2OH \longrightarrow H_2 + O_2 \quad (19)$$
$$HOO + H_2 \longrightarrow H_2O + OH \quad (20)$$
$$HOO + H_2 \longrightarrow H_2O_2 + H \quad (21)$$

There are two chain-propagating sequences (reactions (13) and (14) and reactions (15)–(17)) and one chain-breaking sequence (reactions (18)–(19)). The intermediate is the nitrite as shown in reaction (10). Reaction (11) is the required additional photochemical step. For every NO_2 used to create the O atom of reaction (6), one is formed by reaction (9). However reactions (10), (11), and (15) reveal that for every two molecules of NO consumed, one NO and one NO_2 form; thus the negative quantum yield of NO_2.

With other olefins, other appropriate reactions may be substituted. Ethylene would give

$$O + C_2H_4 \longrightarrow CH_3 + HCO \quad (22)$$
$$HOO + C_2H_4 \longrightarrow 2H_2CO + H \quad (23)$$
$$OH + C_2H_4 \longrightarrow H_2CO + CH_3 \quad (24)$$

Propylene would add

$$OH + C_3H_6 \longrightarrow CH_3CHO + CH_3 \quad (25)$$

Thus PAN would form from

$$CH_3CHO + O_2 \xrightarrow{h\nu} CH_3CO + HOO \quad (26)$$
$$CH_3CO + O_2 \longrightarrow CH_3(CO)OO \quad (27)$$
$$CH_3(CO)OO + NO_2 \longrightarrow CH_3(CO)OONO_2 \quad (28)$$

An acid could form from the overall reaction

$$CH_3(CO)OO + 2CH_3CHO \longrightarrow CH_3(CO)OH + 2CH_3CO + OH \quad (29)$$

Since pollutant concentrations are generally in the parts per million range, it is not difficult to postulate many types of reactions and possible products.

3. The Effect of SO_x

Historically the sulfur oxides were always known to have had a deleterious effect on the atmosphere. Sulfuric acid mist and other sulfate particulate matter have long been recognized as important sources of atmospheric contamination; however, the atmospheric chemistry is probably not as well understood as the gas-phase photoxidation reactions of the nitrogen oxides–hydrocarbon system. The pollutants form originally from the SO_2 emitted to the air. Just as mobile and stationary combustion sources emit some small quantities of NO_2 as well as NO, so do they emit some small quantities of SO_3 when they burn sulfur-containing fuels. Leighton (1961) also discusses the oxidation of SO_2 in polluted atmospheres and more recently an excellent review by Bulfalini (1971) has appeared. This section draws heavily from these sources.

The chemical problem here involves the photochemical and catalytic oxidation of SO_2 and its mixtures with the hydrocarbons and NO, but primarily the concern is with the photochemical reactions, both gas phase and aerosol forming.

The photodissociation of SO_2 into SO and O atoms is quite different from the photodissociation of NO_2. The bond to be broken in the sulfur compound requires 135 kcal/mole. Thus wavelengths greater than 2180 Å do not have sufficient energy to initiate dissociation. This fact is significant in that only solar radiation greater than 2900 Å reaches the lower atmosphere. If there is to be a photochemical effect in the SO_2–O_2 atmospheric system, then it must be that the radiation electrochemically excites the SO_2 molecule but does not dissociate it.

There are two absorption bands of SO_2 within the 3000–4000 Å range. The first is a weak absorption band and corresponds to the transition to the first excited state (a triplet). This band originates at 3880 Å and has a maximum around 3840 Å. The second is a strong absorption band and corresponds to the excitation to the second excited state (a singlet). This band originates at 3376 Å and has a maximum around 2940 Å.

Blacet (1952) carrying out experiments in high O_2 concentrations reported that ozone and SO_3 appeared to be the only products of the photochemically induced reaction. The essential steps were postulated to be

$$SO_2 + h\nu \longrightarrow SO_2^* \tag{30}$$

$$SO_2^* + O_2 \longrightarrow SO_4 \tag{31}$$

$$SO_4 + O_2 \longrightarrow SO_3 + O_3 \tag{32}$$

The radiation used was at 3130 Å, and it would appear that the excited SO_2^* in reaction (30) would be a singlet. The precise roles of the excited singlet and triplet states in the photochemistry of SO_2 are still unclear (Bufalini, 1971). Nevertheless, for purposes here, this point need not be of too great concern since it is possible to write the reaction sequence

$$SO_2 + h\nu \longrightarrow {}^1SO_2^* \tag{33}$$

$${}^1SO_2^* + SO_2 \longrightarrow {}^3SO_2^* + SO_2 \tag{34}$$

Thus reaction (30) could specify either an excited singlet or triplet SO_2^*. The excited state may, of course, degrade by internal transfer to a vibrationally excited ground state which is later deactivated by collision or it may be degraded directly by collisions. Fluorescence of SO_2 has not been observed above 2100 Å. The collisional deactivation steps known to exist in laboratory experiments are not listed here in order to minimize the writing of reaction steps.

Since they involve one specie in large concentrations, reactions (30)–(32) are the primary ones for the photochemical oxidation of SO_2 to SO_3. A secondary reaction route to SO_3 could be

$$SO_4 + SO_2 \longrightarrow 2SO_3 \tag{35}$$

In the presence of water a sulfuric acid mist forms according to

$$H_2O + SO_3 \longrightarrow H_2SO_4 \tag{36}$$

The SO_4 molecule formed by reaction (31) would probably have a peroxy structure and if SO_2^* were a triplet it might be a biradical.

There is conflicting evidence with respect to the results of the photolysis of mixtures of SO_2, NO_x, and O_2. However, many believe that the following should be considered with the NO_x photolysis reactions:

$$SO_2 + NO \longrightarrow SO + NO_2 \tag{37}$$

$$SO_2 + NO_2 \longrightarrow SO_3 + NO \tag{38}$$

$$SO_2 + O + M \longrightarrow SO_3 + M \tag{39}$$

$$SO_2 + O_3 \longrightarrow SO_3 + O_2 \tag{40}$$

$$SO_3 + O \longrightarrow SO_2 + O_2 \tag{41}$$

$$SO_4 + NO \longrightarrow SO_3 + NO_2 \tag{42}$$

$$SO_4 + NO_2 \longrightarrow SO_3 + NO_3 \tag{43}$$

$$SO_4 + O \longrightarrow SO_3 + O_2 \tag{44}$$

$$SO + O + M \longrightarrow SO_2 + M \tag{45}$$

$$SO + O_3 \longrightarrow SO_2 + O_2 \tag{46}$$

$$SO + NO_2 \longrightarrow SO_2 + NO \tag{47}$$

The reducing effect of the SO_2 mentioned in the introduction of this section becomes evident from these reactions.

Some work (Dainton and Irvin, 1950) has been performed on the photochemical reaction between sulfur dioxide and hydrocarbons, both paraffins and olefins. In all cases, mists were found, and these mists settled out in the reaction vessels as oils which had the characteristics of sulfinic acids. Because of the small amounts of materials formed there are great problems in elucidating particular steps. When NO_x and O_2 are added to this system, the situation is most complex. Bufalini (1971) sums up the status in this way, ". . . work indicates that the aerosol formed from mixtures of the lower hydrocarbons with NO_x and SO_2 is predominantly sulfuric acid, whereas the higher olefin hydrocarbons appear to produce carbonaceous aerosols also, possibly organic acids, sulfinic or sulfonic acids, nitrate-esters, etc."

B. NO_x FORMATION AND REDUCTION

The previous section established the great importance of NO_x in the photochemical smog reaction cycles. Strong evidence exists that the major producer of NO_x has been the automobile. But as automobile emissions' standards are enforced and electricity-generating plants turn from natural gas to coal and oil as fuels, there is no doubt that stationary sources will contribute a heavier fraction of the total NO_x emitted to the atmosphere. Consequently there has been great interest in predicting NO_x emissions and this interest has led to the formulation of various analytical models to predict specifically NO formation in combustion systems.

The greatest number of analytical and experimental studies have been focused on NO formation alone and not on NO_2. Indeed the major portion of NO_x has been found to be NO. Recent measurements in practical combustion systems, particularly those used to simulate aircraft gas turbine conditions, have shown larger amounts of NO_2 than one would expect. Controversy surrounds this question of NO_2 formation, and many believe that the NO_2 measured in some experiments actually formed in the probe systems used to remove a sample of gas.

Another controversy which has made the study of the NO_x chemistry confusing revolved around the question of "prompt" NO postulated to form in the flame zone by mechanisms other than those thought to hold exclusively for NO formation from atmospheric nitrogen. Although the amount of NO formed by the so-called "prompt" condition is usually very small, the fundamental studies into this problem have helped clarify much about NO_x formation both from atmospheric and fuel-bound nitrogen.

NO formation from atmospheric nitrogen is meant to specify the NO formed in combustion systems in which the original fuel contains no nitrogen atoms which are chemically bound to other elements such as carbon and hydrogen. Generally it is thought to be the NO formed from the nitrogen in air. Since this NO forms only at high temperatures, it is sometimes called the thermal NO. The "prompt" NO work has shown, however, that in the flame zone the nitrogen in the air can form small quantities of CN compounds which are subsequently oxidized to NO. The stable compound HCN has been found as a product in very fuel-rich flames.

NO_x formation from fuel-bound nitrogen is meant to specify the NO formed from fuel compounds that contain nitrogen atoms bound to other atoms. Generally these nitrogen atoms are bound to carbon or hydrogen atoms. Fuel-bound nitrogen compounds are ammonia, pyridine, and many other amines. The amines can be designated as the other organic compounds have in Chapter 3 as $R{-}NH_2$, where R is an organic radical or H atom. The NO formed from HCN and the fuel fragments from the nitrogen compounds are sometimes referred to as chemical NO in an analogous terminology to the thermal NO mentioned in the last paragraph.

1. The Structure of the Nitrogen Oxides

Many investigators have attempted to investigate analytically the formation of NO in fuel–air systems. Because of the availability of an enormous computer capacity, they have written all the possible reactions of the nitrogen oxides which they thought possible. Unfortunately some of these investigators have ignored the fact that some of the reactions written could have been eliminated because of steric considerations. Since the structure of the various nitrogen oxides can be important, their formulas and structures are given in Table 1.

2. The Effect of Flame Structure

As the important effect of temperature on NO formation is discussed in the following sections, it is well to remember that flame structure can play a most significant role in determining the overall NO_x emitted. For premixed systems such as that obtained on Bunsen and flat flame burners and almost obtained in carbureted spark-ignition engines, the temperature, and thus the mixture ratio, is the prime parameter in determining the quantities of NO_x formed. Ideally, as in equilibrium systems,

TABLE 1

Structure of gaseous nitrogen compounds

Compound	Structure	
Nitrogen N_2	$N{\equiv}N$	
Nitrous oxide N_2O	$^-N{=}N^+{=}O$	$N{\equiv}N^+{-}O^-$
Nitric oxide NO	$N{=}O$	
Nitrogen dioxide NO_2		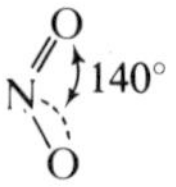 O=N–O (140°)
Nitrate radical NO_3		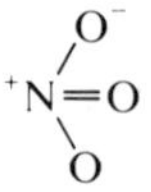 $^+N(=O)(-O^-)(-O)$
Nitrogen tetroxide N_2O_4		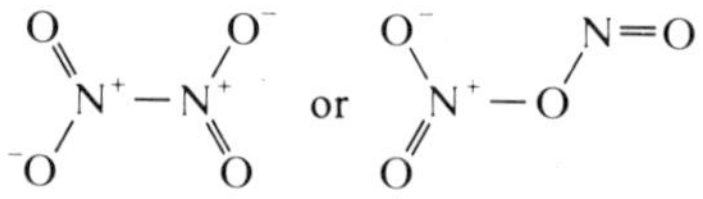 $(O{=})(^-O{-})N^+{-}N^+(-O^-)(=O)$ or $(O^-{-})(O{=})N^+{-}O{-}N{=}O$
Nitrogen pentoxide		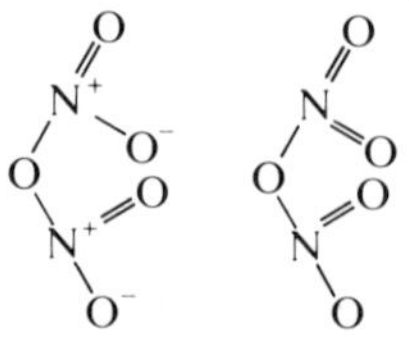 $O{=}N^+(-O^-){-}O{-}N^+(=O)(-O^-)$; $O{=}N(=O){-}O{-}N(=O){-}O$

the NO formation should peak at the stoichiometric value and decline on both the fuel-rich and fuel-lean sides, just as the temperature does. Actually because of kinetic (nonequilibrium) effects, the peak is found somewhat on the lean (oxygen-rich) side of stoichiometric.

However in fuel injection systems, even though the overall mixture ratio may be lean and the final temperature could correspond to this overall mixture ratio, the fuel droplets or fuel jets burn as diffusion flames. The temperature of these diffusion flames are at the stoichiometric values during part of the burning time, even though the excess species will eventually dilute the products of the flame to reach the true equilibrium final temperature. Thus in diffusion flames, more NO_x forms than would be expected from a calculation of an equilibrium temperature based on the overall mixture ratio. The reduction reactions of NO are so slow that in most practical systems the amount of NO formed in diffusion flames is unaffected by the subsequent drop in temperature caused by dilution of the excess species.

3. Atmospheric Nitrogen Kinetics

For premixed systems a conservative estimate can be made of the NO formation from consideration of the equilibrium given by reaction (48).

$$N_2 + O_2 \rightleftharpoons 2NO \tag{48}$$

However, the kinetic route of NO formation is not the attack of an oxygen molecule on a nitrogen molecule. Mechanistically, oxygen atoms are formed from the dissociation of O_2 or from an H atom attack on O_2 and these oxygen atoms attack nitrogen molecules to start the simple chain shown by reactions (49) and (50).

$$O + N_2 \longrightarrow NO + N, \quad k = 1.4 \times 10^{14} \exp(-78{,}500/RT) \tag{49}$$

$$N + O_2 \longrightarrow NO + O, \quad k = 6.4 \times 10^{9} \exp(-6280/RT) \tag{50}$$

This chain was first postulated by Zeldovich (1946) and is referred to as the Zeldovich mechanism. It is thought to be the main source of NO in most combustion systems. However, emission standards on NO_x have been so stringent that some have included the reaction

$$N + OH \longrightarrow NO + H, \quad k = 2.8 \times 10^{13} \tag{51}$$

in the overall atmospheric nitrogen oxidation system. The contribution of this reaction must be small because the reacting species are both radicals which are therefore present only in very small quantities.

The high activation energy reaction (49) obviously controls the system and it is apparent then that the O radical concentration becomes critical in the system. To prevent NO formation in air systems, one must prevent O atom formation.

Because reaction (49) is so slow due to its high activation energy, many believed that in premixed flame systems the NO would form only in the postflame or burned gas zone. Thus it was thought that it would be possible to determine the rate of formation and the rate constant of NO by measurements of the NO concentration profiles past the flame in the postflame zone. Such measurements can be obtained readily on flat flame burners. In order to make these determinations, it is necessary to know the O atom concentrations. The nitrogen concentration was always in large excess and known. The O atom concentration was taken as the equilibrium concentration at the flame temperature. The thought is that all other reactions are very fast compared to the Zeldovich mechanism.

Experimental measurements on flat flame burners showed that when the NO concentration profiles were extrapolated to the flame front position the concentration of NO does not go to zero, but some finite value. Such results were most prevalent with fuel-rich flames. Fenimore (1971) argued that reactions other than the Zeldovich mechanism were playing a role in the flame and that some NO was being formed in the flame region. He called this NO, "prompt" NO. He noted that "prompt" NO was not found in nonhydrocarbon flames such as CO–air and H_2–air, which were analyzed experimentally in the same manner as the hydrocarbon flames. The reaction scheme he suggested to explain the NO found in the flame zone involved a hydrocarbon species and atmospheric nitrogen. The nitrogen compound formed reacted with O radicals via the following mechanism

$$\begin{aligned} CH + N_2 &\rightleftharpoons HCN + N \\ C_2 + N_2 &\rightleftharpoons 2CN \end{aligned} \tag{52}$$

The N atoms could form NO, in part at least, by reactions (50) and (51), and the CN could yield NO by oxygen or oxygen atom attack. It is well known that CH exists in flames and indeed, as stated in Chapter 4, is the molecule which gives the deep violet color to a Bunsen flame.

In order to verify whether reactions other than the Zeldovich mechanism were effective in NO formation, various investigators undertook the study of the CH_4–O_2–N_2 reacting system at a fixed high temperature and work in this area was that of Bowman and Seery (1972) who studied the CH_4–O_2–N_2 system. Martenay (1970), a colleague of Bowman and Seery, showed that it was possible to perform the complex kinetic calculations of the CH_4—O_2—N_2 reacting system at a fixed high temperature and pressure similar to those obtained in a shock tube. Martenay's results for $T = 2477°K$ and $P = 10$ atm are shown in Fig. 1. These results are worth considering in their own right for they show explicitly much that has been inferred. Examination of Fig. 1 shows that at about 5×10^{-5} sec, all the energy release reactions have equilibrated before any significant amounts

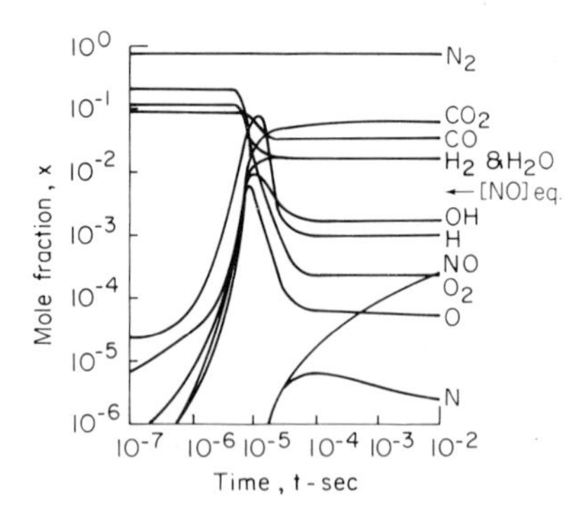

Fig. 1. Concentration–time profiles in kinetic calculation of methane–air reaction with inlet temperature of 1000°K. $P_{in} = 10$ atm, $\phi = 1.0$, and $T_c = 2477°K$ (after Martenay, 1970).

of NO have formed and, indeed, even at 10^{-2} sec the NO has not reached its equilibrium concentration for $T = 2477°K$. These results show that for such homogeneous, or near homogeneous, reacting systems it would be possible to quench the NO reactions, obtain the chemical heat release and prevent NO formation. In certain combustion schemes, this procedure has been put in practice.

Figure 1 also shows a large oxygen radical overshoot within the reaction zone. It was thought that, if within the reaction zone, the O atom concentration could be orders of magnitude greater than its equilibrium value, then this condition could lead to the "prompt" NO found in flames. The mechanism analyzed to obtain the results depicted in Fig. 1 was essentially that given in Section 3.D.3a with the Zeldovich reactions. Thus it was thought possible that the Zeldovich mechanism could account for the "prompt" NO.

The experiments of Bowman and Seery appeared to confirm this conclusion. Some of Bowman's results are shown in Fig. 2. In this figure

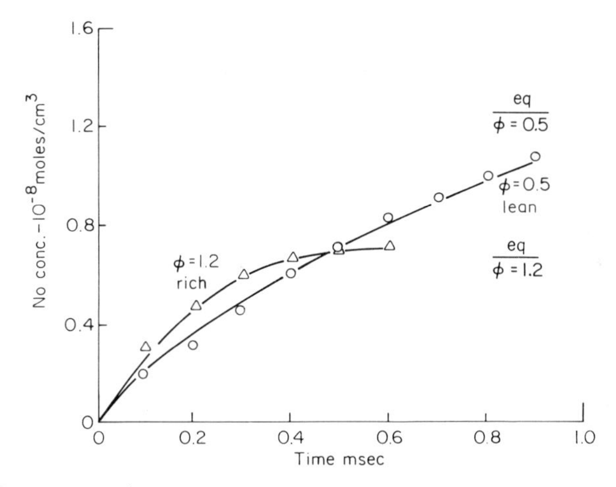

Fig. 2. Comparison of measured and calculated NO concentration profiles for CH_4–O_2–N_2–Ar mixtures behind reflected shocks. Initial post shock conditions $T = 2960°K$, $P = 3.2$ atm (after Bowman, 1973).

the experimental points correlate very well with the analytical calculations based on the Zeldovich mechanisms alone. Bowman and Seery used the same computational program as Martenay. Figure 2 also depicts another frequent result which is that fuel-rich systems approach NO equilibrium much faster than fuel-lean systems.

Although Bowman and Seery's results would seem to refute the suggestion by Fenimore that "prompt" NO forms by reactions other than the

Zeldovich mechanism, one must remember that flames and shock tube initiated reacting systems are distinctively different processes. In a flame there is a temperature profile which begins at the ambient temperature and proceeds to the flame temperature. Thus although flame temperatures may be simulated in shock tubes, the reactions in flames are initiated at much lower temperatures than those in shock tubes. As stressed many times before, the temperature history frequently determines the kinetic route and the products. Thus the shock tube results do not prove that the Zeldovich mechanism alone determines NO formation. The "prompt" NO could arise from other reactions in flames as suggested by Fenimore.

Bachmeier *et al.* (1973) appear to substantiate Fenimore's postulates and to give greater insight to the flame NO problem. They measured the "prompt" NO formed as a function of equivalence ratio for many hydrocarbon compounds. These results are shown in Fig. 3. What is significant about these results is that the maximum "prompt" NO is reached on the fuel-rich side of stoichiometric, remains at a high level through a fuel-rich region, and then drops off sharply about an equivalence ratio of 1.4.

Bachmeier *et al.* also measured the HCN concentrations through

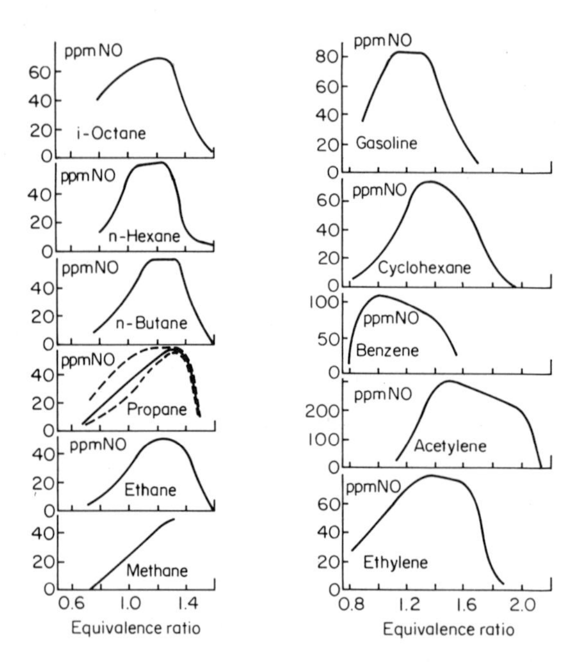

Fig. 3. "Prompt NO" as a function of mixture strength and fuel. The dotted lines show the uncertainty of the extrapolation at the determination of "prompt NO" in propane flames; similar curves were obtained for the other hydrocarbons (after Bachmeier *et al.*, 1973).

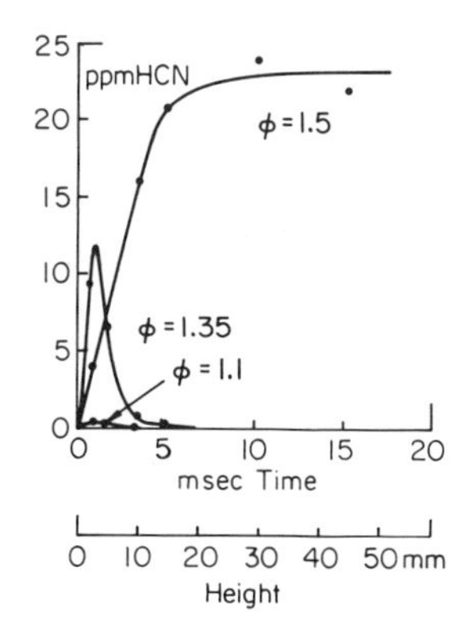

Fig. 4. HCN profiles of fuel-rich propane–air flames (after Bachmeier *et al.*, 1973).

propane–air flames. These results, which are shown in Fig. 4, show that HCN concentrations rise sharply somewhere in the flame, reach a maximum and then decrease sharply. However, for an equivalence ratio of 1.5, a fuel-rich condition for which little "prompt" NO is found, the HCN continues to rise and is not depleted. The explanation offered for this trend is that HCN forms in all the rich hydrocarbon flames; however, below an equivalence ratio of 1.4 there are still sufficient O radicals present to react with HCN to deplete it and to form the NO. Since the sampling and analysis techniques used by Bachmeier *et al.* (1973) would not permit the identity of CN, the cyanogen radical, the HCN concentrations probably represent the sum of CN and HCN as they exist in the flame. The CN and HCN in the flame are related through the rapid equilibrium reactions (Haynes *et al.*, 1975)

$$CN + H_2 \rightleftharpoons HCN + H$$

$$CN + H_2O \rightleftharpoons HCN + OH$$

The HCN concentration is most likely reduced mainly by the oxidation of the CN radicals (Haynes *et al.*, 1975; Leonard *et al.*, 1976).

From other more recent studies of NO formation in the combustion of lean and slightly rich methane–oxygen–nitrogen mixtures and lean and very rich hydrocarbon–oxygen–nitrogen mixtures, it must be concluded that some of the prompt NO is due to the overshoot of O and OH radicals above their equilibrium values, as Bowman and Seery initially thought. However, even though there can be O radical overshoot on the fuel-rich side of stoichiometric, this overshoot cannot explain the "prompt" NO formation in fuel-rich systems. It would appear that both the Zeldovich and Fenimore mechanisms could hold. Some very interesting experiments by

Eberius and Just (1973) would seem to clarify what is happening in the flame zone with regard to NO formation.

Eberius and Just's experiments were performed on a flat flame burner with propane as the fuel. Measurements were made of the "prompt" NO at various fuel–oxygen equivalence ratios whose flame temperatures were controlled by dilution with nitrogen. Thus a range of temperatures could be obtained for a given propane–oxygen equivalence ratio. The results obtained are shown in Fig. 5. The highest temperature point for each equivalence ratio corresponds to zero dilution.

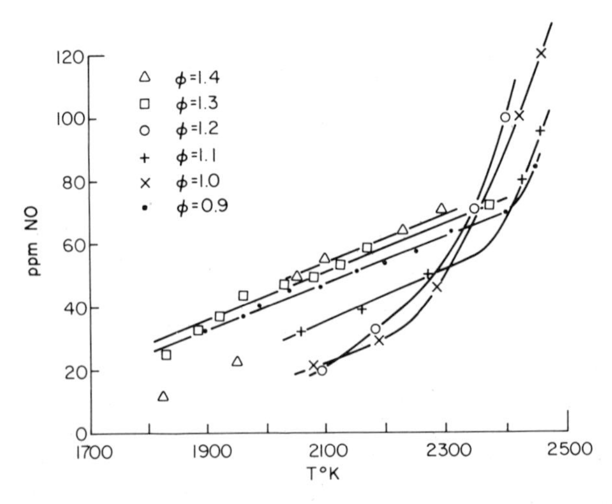

Fig. 5. "Prompt NO" as a function of the temperature at various mixture strengths Φ in adiabatic propane–synthetic air flames (after Eberius and Just, 1973).

The shape of the plots in Fig. 5 are revealing. At both the low and higher temperature ends, all plots seem nearly parallel. The slopes at the low temperature end are very much less than the slopes at the high temperature end and would indicate that there are indeed two mechanisms for the formation of "prompt" NO. The two mechanisms are not related solely to the fuel-rich and fuel-lean stoichiometry as many investigators thought, but to the flame temperature. These results indicate that there is a high temperature, high activation route and a lower temperature, low activation route.

The systematic appearance of these data led Eberius and Just to estimate the activation energy for the two regions. Without correcting for diffusion, they obtained an activation energy of the order of 65 kcal/mole for the high temperature zone. This value is remarkably close to the 75 kcal/mole activation energy for the initiating step in the Zeldovich mechanism. Further, diffusion corrections would raise the experimental value somewhat. The low temperature region has an activation energy of the order of 12–16 kcal/mole. As will be shown later, radical attack on the cyano species is faster than oxygen radical attack on hydrogen. The activation energy of $O + H_2$ is about 8 kcal/mole, thus the HCN reaction should be less. Again, diffusion corrections for the oxygen atom concentration could lower the apparent activities of 12–16 kcal/mole to below 8 kcal/mole. It would appear that even this crude estimate of the activation energy from Eberius and Just's low temperature region and the formation of HCN found by the same group (Bachmeier *et al.*, 1973) in their other flame studies with propane (Fig. 4) would indicate that the Fenimore mechanism would hold in the lower temperature region.

To conclude, it would appear that "prompt" NO could form in the flame zone by either the Zeldovich mechanism or the Fenimore mechanism, depending upon the flame temperature and stoichiometry. It is to be emphasized that the "prompt" NO is only a small fraction of the NO that forms in the post flame zone where the Zeldovich mechanism controls completely.

4. Fuel-Bound Nitrogen Kinetics

In several recent experiments, it has been shown that NO emissions from combustion devices that operated with nitrogen containing compounds in the fuel were high or, in other words, fuel-bound nitrogen is an important source of NO. The initial experiments of Martin and Berkau (1971) commanded the greatest interest. These investigators added 0.5% pyridine to base oil and found almost an order of magnitude increase over the NO formed from base oil alone. Their results are shown in Fig. 6.

During the combustion of fuels containing bound nitrogen compounds, the nitrogen compounds will most likely undergo some thermal decomposition prior to entering the combustion zone. Hence, the precursors to NO formation will, in general, be low molecular weight nitrogen-containing compounds or radicals (NH_2, HCN, CN, NH_3, etc.). All indications are that the oxidation of fuel-bound nitrogen compounds to NO is rapid and occurs on a time scale comparable to the energy release reactions in the combustion systems. In fact, in the vicinity of the combustion zone, observed NO concentrations significantly exceed calculated equilibrium values. In the postcombustion zone, the NO concentration decreases relatively slowly for fuel-lean mixtures and more rapidly for fuel-rich mixtures. Recall Bowman and Seery's results (Fig. 2) showing that fuel-rich systems approach equilibrium faster. When fuel nitrogen compounds are present, high NO yields are obtained for lean and stoichiometric mixtures and relatively lower

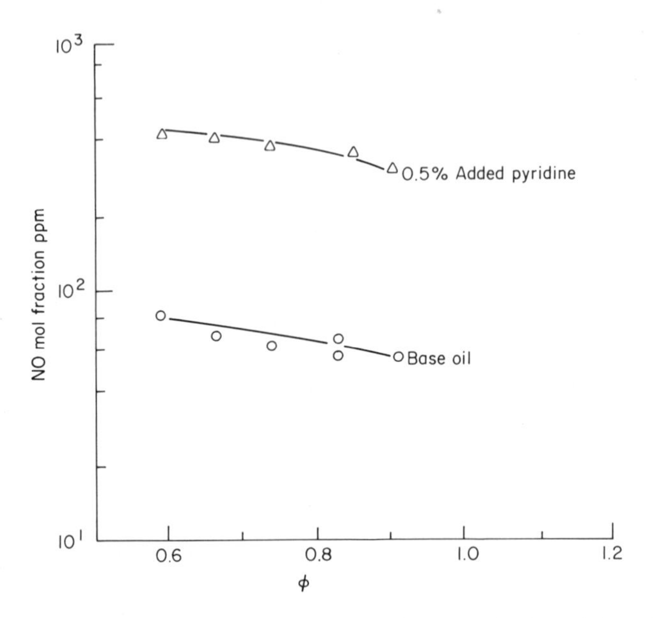

Fig. 6. Nitric Oxide emissions from oil-fired laboratory furnace (after Martin and Berhaus, 1971).

yields are found for fuel-rich mixtures. The NO yields appear to be only slightly dependent on temperature and thus indicate a low activation energy step. This result should be compared to the strong temperature dependence of NO formation from atmospheric nitrogen.

The high yields on the lean side of stoichiometric pose a dilemma. It is desirable to operate lean to reduce hydrocarbon and carbon monoxide emissions, but with fuel containing bound nitrogen high NO yields would be obtained. The reason for the superequilibrium yields is that the reactions leading to the reduction of NO to its equilibrium concentration, namely,

$$\mathrm{O + NO \longrightarrow N + O_2} \tag{53}$$

$$\mathrm{NO + NO \longrightarrow N_2O + O} \tag{54}$$

$$\mathrm{NO + RH \longrightarrow products} \tag{55}$$

are very slow. NO can be reduced under certain conditions by CH and NH radicals which can be present in relatively large concentrations in fuel-rich systems. These reduction steps and their application will be discussed later.

The characteristic reactions by which NO forms from fuel-bound nitrogen compounds are written in the form of the fragments which form from pyrolysis of the original compounds early in the combustion process. A phenomenological scheme would be

$$\left.\begin{matrix} NH \\ \\ CN \end{matrix}\right\} + \left\{\begin{matrix} O_2 \\ \text{or} \\ 2O \end{matrix}\right\} \longrightarrow \begin{matrix} NO + OH \\ \\ CO + NO \end{matrix} \tag{56}$$

A reasonable qualitative conversion route for the NH could be (Flagan *et al.*, 1974)

$$NH + O \longrightarrow NO + H$$

$$NH + O \longrightarrow N + OH$$

$$N + O_2 \longrightarrow NO + O$$

$$N + OH \longrightarrow NO + H$$

All the reactions are very exothermic and spin allowed. The details of the cyanogen conversion mechanism have been reported by Mulvihill and Phillips (1975) and essentially appear to follow the route

$$CN + O_2 \longrightarrow OCN + O$$

$$OCN + O \longrightarrow CO + NO$$

which is also quite exothermic. The reaction of HCN with O radicals

$$HCN + O \longrightarrow HC + NO \tag{57}$$

as with all other oxidizing species is very endothermic and is not likely to proceed in competition with the other oxidation steps.

Reactions (56) are faster (Flagan *et al.*, 1974; Mulvihill and Phillips, 1975) than the competing reaction

$$O + H_2 \longrightarrow OH + H \tag{58}$$

and provide the reason for the earlier statement that the NO formation reactions are of the same order as the energy release reactions in the combustion system. Thus it is not possible to quench the reaction system to prevent NO formation from fuel-bound nitrogen as it is for atmospheric nitrogen.

The mechanism for the NO formation here is much like the Fenimore mechanism for "prompt" NO formation. The nitrogen fragment in the Fenimore mechanism can form not only from the reaction

$$CH + N_2 \longrightarrow HCN + N \tag{59}$$

mentioned previously, but also from

$$C_2 + N_2 \longrightarrow 2CN \tag{60}$$

$$C + N_2 \longrightarrow CN + N \tag{61}$$

Again the green color of fuel-rich Bunsen flames is due to the presence of the C_2 radicals which are known to exist in these flames. Carbon vapor must also be present as C atoms in small concentrations. It is very difficult to identify C atoms spectroscopically. Only metal atoms with low ionization potentials radiate in the visible.

In fuel-rich systems, there is evidence (Fenimore, 1972; de Soete, 1975; Bowman, 1975) that the fuel nitrogen intermediate reacts not only with oxidizing species in the manner represented by reactions (56) but also in a competitive manner with NO (or another nitrogen intermediate) to form N_2. This second step, of course, is the reason that there are lower NO yields in fuel-rich systems. The fraction of fuel nitrogen converted to NO in fuel-rich systems can be as much as an order of magnitude less than that of lean or near-stoichiometric systems. One should realize, however, that even in fuel-rich systems the exhaust NO concentration is substantially greater than its equilibrium value at the combustion temperature.

Haynes *et al.* (1975) have shown that when small amounts of pyridine are added to a premixed rich ($\phi = 1.68$, $T = 2030°K$) ethene–air flame, the amount of NO increases with little decay of NO in the post-flame gases. However, when larger amounts of the pyridine are added, significant decay

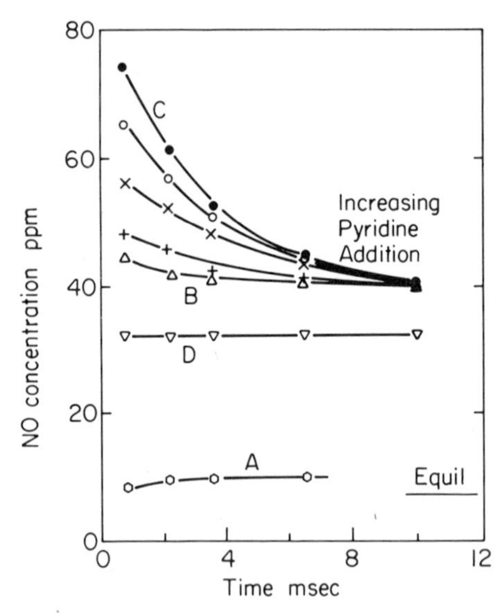

Fig. 6A. Effect on NO concentrations leaving the reaction zone of an ethylene–air flame ($\phi = 1.68$, $T = 203°L$) with various pyridine additions. Curve A, no pyridine addition; Curves B and C, 0.1–0.5% N by weight of fuel; Curve D, NO addition to fuel–air mixture (after Hayes *et al.*, 1975).

of NO is observed after the reaction zone. Increasingly higher pyridine additions result in high NO concentrations leaving the reaction zone, but this concentration drops appreciably in the post-flame gases to a value characteristic of the flame, *but well above the calculated equilibrium value.* Actual experimental results are shown in Fig. 6A.

In fuel-rich systems, the conversion reactions of the fuel nitrogen intermediates must be considered in doubt, mainly because the normal oxidizing species O_2, O, and OH are present only in very small concentrations, particularly near the end of the reaction zone. Haynes *et al.* (1975) offer the interesting suggestion that the CN can be oxidized by CO_2 since the reaction

$$CN + CO_2 \longrightarrow OCN + CN$$

is 20 kcal/mole exothermic and is estimated to be reasonably fast.

5. The Formation of NO_2

In recent emission sampling studies, significant NO_2 concentrations were measured in gas turbine exhausts, and *in situ* measurements of nitrogen oxide concentrations in gaseous turbulent diffusion flames indicate that there are relatively large NO_2/NO ratios near the combustion zone. These results are surprising in that chemical equilibrium considerations reveal that, for typical flame temperatures, the NO_2/NO ratios should be negligibly small. Further, kinetic models, such as those mentioned when modified to include reactions for NO_2 formation and reduction, show that in practical devices, the conversion of NO to NO_2 can be neglected.

These anomalies led Merryman and Levy (1974) to a closer examination of both NO and NO_2 formation in a flat flame burner operated near stoichiometric. Their measurements show that NO_2 is produced in the visible regime of all air flames (with and without fuel-bound nitrogen) and that NO is only observed in the visible when there is fuel-bound nitrogen. Further, these investigators found that the NO_2 is consumed rapidly in the near-post flame zone and the NO concentration rises correspondingly. They postulated the following scheme to represent their findings:

$$\left.\begin{matrix} NH \\ CN \end{matrix}\right\} + O_2 \longrightarrow NO + \cdots \qquad (62)$$

$$NO + HO_2 \longrightarrow NO_2 + OH \qquad (63)$$

$$NO_2 + O \longrightarrow NO + O_2 \qquad (64)$$

The significant step is represented by reaction (63). One should recall that in the early parts of a flame there can be appreciable amounts of HO_2. The appearance of the NO_2 is supported further by the fact that reaction

(63) is two orders of magnitude faster than reaction (64). The importance of the hydroperoxy radical attack on NO was verified further by adding NO to the cold fuel mixtures in some experiments. In these tests the NO disappeared before the visible region was reached in oxygen-rich and stoichiometric flames, i.e., flames that would produce HO_2. The NO_2 persists because, as mentioned previously, its reduction to N_2 and O_2 is very slow. The role of HO_2 would not normally be observed in shock tube experiments because of the high temperatures which exist.

The Merryman–Levy sequence could explain the experimental results which show high NO_2/NO ratios. In the experiments in which these high ratios were found, it is quite possible that reaction (64) is quenched and the NO_2 is not reduced.

6. The Reduction of NO_x

Because of the stringent emission standards imposed on both mobile and stationary power sources, methods for reducing NO_x must be found, and these methods should not impair the efficiency of the device. The simplest method of reducing NO_x, particularly from gas turbines, is by water addition to the combustor can. Water vapor can reduce the O radical concentration by the following scavenging reaction

$$H_2O + O \longrightarrow 2OH \tag{65}$$

Fortunately OH radicals cannot attack N_2. However it is more likely that the effect of water on NO_x emissions is through the attendant reduction in combustion temperature. NO_x formation from atmospheric nitrogen arises primarily from the very temperature sensitive Zeldovich mechanism.

In heterogeneous systems such as those which arise due to direct liquid fuel injection and which are known to burn as diffusion flames, the problem of NO_x reduction is more difficult. One possible means is to decrease the average droplet size formed from injection. Kesten (1971) and Bracco (1973) have shown that the amount of NO formed from droplet diffusion flames can be related to the droplet size, in that a large droplet will give more NO than that obtained from a group of smaller droplets equal to the mass of the larger droplet. Any means of decreasing the heterogeneity of a flame system will decrease the NO_x. Another possible practical scheme is to emulsify the fuel with a higher vapor pressure, nonsoluble component such as water. It has been shown (Dryer, 1974) that droplets from such emulsified fuels explode after combustion has been initiated. These microexplosions occur when the superheated water within the fuel droplet vaporizes and appreciably decreases the heterogenity of the system. A further benefit is obtained by not only having water present for dilution, but also having the water present in the intimate vicinity of the diffusion flame.

If it is impossible to reduce the amount of NO_x in the combustion section of a device, then the NO_x must be removed somewhere in the exhaust. Myerson (1974) has shown that it is possible to reduce NO_x by adding small concentrations of fuel and oxygen. The addition of about 0.1% hydrocarbon (isobutane) and 0.4% O_2 to a NO_x-containing system at 1260°K reduced the NO_x concentration by a factor of two in about 125 msec. Myerson found that the ratio of O_2/HC was most important. At large concentrations of O_2 and the hydrocarbon, an HCN formation problem could arise. This procedure will only hold for slightly fuel-lean or fuel-rich systems. The oxygen is the creator and the destroyer of the species involved in the NO reduction. This fact in turn means that the initial addition of O_2 to the hydrocarbon–NO mixture promotes the production of the strongly reducing species CH and CH_2 and similar substituted free radicals which otherwise must be produced by slower pyrolysis reactions.

Continued addition of O_2 beyond one-half the stoichiometric value with the hydrocarbons present encourages a net destruction of the hydrocarbon radicals. For the temperature range 1200–1300°K, production of the hydrocarbon radicals via hydrogen abstraction by O_2 is rapid, even assuming an activation energy of 45 kcal/mole, and more than adequate to provide sufficient radicals for NO reduction in the stay time range of 125 msec.

The reactions postulated by Myerson to be involved are

$$CH + NO \longrightarrow HCO + N + 52 \quad kcal \tag{66}$$

$$CH + NO \longrightarrow HCN + O + 73 \quad kcal \tag{67}$$

The exothermicity of reaction (66) is sufficient to fragment the formyl radical and could be written as

$$CH + NO \longrightarrow H + CO + N + 25 \quad kcal \tag{68}$$

The N radicals in the absence of O_2 in these fuel-rich systems can react rapidly with NO via

$$N + NO \longrightarrow N_2 + O + 73 \quad kcal \tag{69}$$

Others have proposed to carry out similar homogeneous catalytic reduction schemes by the appropriate addition of ammonia and oxygen instead of a hydrocarbon and oxygen.

7. The Partial Equilibrium Assumption

In many of the experiments discussed, in the previous sections of this chapter, it was necessary to know the O atom concentrations in order to determine the NO formation rates. It is difficult to measure the concentration of atomic species; however, one can readily measure the concentration of diatomic radicals such as OH by infrared spectroscopic means and the other stable species present. In order to determine the concentration of O radicals, it is assumed that the rapid bimolecular reactions

$$H + O_2 \rightleftarrows OH + O \tag{70}$$

$$O + H_2 \rightleftarrows OH + H \tag{71}$$

$$H_2 + OH \rightleftarrows H + H_2O \tag{72}$$

are locally equilibrated even though there is not equilibration of the other species in the reacting system.

The equilibrium constants for reactions (70)–(72) are

$$K_{P70} = (OH)(O)/(H)(O_2) \tag{73}$$

$$K_{P71} = (OH)(H)/(O)(H_2) \tag{74}$$

$$K_{P72} = (H_2O)(H)/(OH)(H_2) \tag{75}$$

Eliminating the H radical concentration from Eqs. (74) and (75), one obtains

$$K_{P71} = [(OH)/(O)][K_{P72}(OH)/(H_2O)] \tag{76}$$

or

$$(O) = [(OH)^2/(H_2O)][K_{P72}/K_{P71}] \tag{77}$$

The H radical concentration is then found to be

$$(H) = K_{P72}(OH)(H_2)/(H_2O) \tag{78}$$

The equilibrium constant can be calculated and the OH and stable species measured.

Indeed, if one cannot make spectroscopic measurements to determine the (OH) concentration, then it is possible to relate this concentration to the stable species in the system. As discussed in Chapter 3, the rate of formation of CO_2 from wet CO oxidation is via the reaction

$$CO + OH \longrightarrow CO_2 + H \tag{79}$$

Thus the rate expression is

$$d(CO_2)/dt = -d(CO)/dt = k(CO)(OH) \tag{80}$$

Now if one again assumes equilibrium in the H_2–O_2 chain that must exist in the wet CO case, he can write the following equilibrium reactions of formation

$$\tfrac{1}{2}H_2 + \tfrac{1}{2}O_2 \rightleftarrows OH \tag{81}$$

$$H_2 + \tfrac{1}{2}O_2 \rightleftarrows H_2O \tag{82}$$

with the equilibrium constants

$$K^2_{P, f\text{OH}} = (\text{OH})^2/(\text{H}_2)(\text{O}_2) \tag{83}$$

$$K_{P, f\text{H}_2\text{O}} = (\text{H}_2\text{O})/(\text{H}_2)(\text{O}_2)^{1/2} \tag{84}$$

If the H_2 concentration is eliminated from Eqs. (83) and (84), an expression for the OH concentration can be obtained in terms of the H_2O and O_2 concentrations, i.e.,

$$(\text{OH})_{\text{eq}} = (\text{H}_2\text{O})^{1/2}\,(\text{O}_2)^{1/4}[K^2_{P, f\text{OH}}/K_{P, f\text{H}_2\text{O}}]^{1/2} \tag{85}$$

Substituting this value in the rate expression (reaction (70)) gives

$$d(\text{CO}_2)/dt = -d(\text{CO})/dt = k_{79}[K^2_{P, f\text{OH}}/K_{P, f\text{H}_2\text{O}}]^{1/2}(\text{CO})(\text{H}_2\text{O})^{1/2}(\text{O}_2)^{1/4} \tag{86}$$

Thus it is seen that in CO oxidation the rate expression can be written in terms of the stable species when the partial equilibrium assumption is made. Care must be exercised in applying this assumption. Such equilibria do not always exist in systems and, indeed, it is in question whether they exist in the pure, wet CO oxidation system.

C. SO$_x$ EMISSIONS

Sulphur compounds pose a dual problem. Not only do their combustion products contribute to atmospheric pollution, as described in a previous section, but these products are also very corrosive in nature and cause severe physical problems in gas turbines and industrial power plants. Sulfur pollution and corrosion were problems long before the nitrogen oxides were known to affect the atmosphere. However, the general availability of low sulfur fuels diminished the concern with respect to the sulfur. Now that the availability of low sulfur crude oil is very restricted, greater attention is being given to the use of coal and those crudes which have "appreciable" sulfur content. Costs for removing sulfur from residual oils by catalytic hydrodesulfurization techniques remain high, and the desulfurized residual oil have a tendency to become "waxy" at low temperatures. To remove sulfur from coal is an even more imposing problem. It is possible to remove pyrites from coal, but this approach is limited by the size of the pyrite particles. Unfortunately, pyrite sulfur makes up only half the sulfur content of coal, whereas the other half is organically bound. Coal gasification is the only means by which this sulfur mode could be removed. Of course, it is always possible to eliminate the deleterious effects of sulfur by removing the major product oxide SO_2 by absorption processes.

These processes impose large initial capital investments and most industries believe these are too costly for implementation.

The presence of sulfur compounds in the combustion process can affect the nitrogen oxides as well. Thus, a study of sulfur compound oxidation is not only important from the point of view of possibly offering alternate or new means of controlling the emission of the objectionable sulfur oxide, but also with regard to their effect on the formation and concentration of other pollutants.

There are some very basic differences between the sulfur problem and that posed by the formation of the nitrogen oxides. The two possible sources of nitrogen in any combustion process are either atmospheric or organically bound. Sulfur can be present in elemental form, organically bound, or as a species in various inorganic compounds. Once it enters the combustion process it is very reactive with oxidizing species, and similar to fuel nitrogen, its conversion to the sulfurous oxides is fast compared to the other energy releasing reactions.

Even though sulfur oxides posed a problem in combustion processes well before the concern for photochemical smog and the role of the nitrogen oxides in creating this smog, much less is understood about the mechanisms of sulfur oxidation. Indeed the amount of recent work on sulfur oxidation has been minimal. The status of the field has been reviewed by Levy *et al.* (1970), and Cullis and Mulcahy (1972), and much of the material from the following subsections has been drawn from Cullis and Mulcahy's article.

1. The Product Composition and Structure of Sulfur Compounds

In any combustion system in which elemental sulfur or a sulfur bearing compound is present, the predominant product is sulfur dioxide. The concentration of sulfur trioxide found in combustion systems is most interesting. Even under very lean conditions, the amount of sulfur trioxide formed is only a few percent of that of sulfur dioxide. However, generally the sulfur trioxide concentration is higher than its equilibrium value, as would be expected from the reactions

$$SO_2 + \tfrac{1}{2}O_2 \rightleftharpoons SO_3 \tag{87}$$

These higher than equilibrium concentrations appear to be due to the fact that the homogeneous reactions which would reduce the SO_3 to SO_2 and O_2 are known to be slow. This point will be discussed later in this section.

It is well known that SO_3 has a great affinity for water and that at low temperatures it appears as sulfuric acid. Above 500°C, sulfuric acid dissociates almost completely into sulfur trioxide and water.

Under fuel-rich conditions, in addition to sulfur dioxide the stable products are found to be hydrogen sulfide, carbonyl sulfide, and elemental sulfur.

There are other oxides of sulfur which, due to their reactivity, appear only as intermediates in various oxidation reactions. These are sulfur monoxide SO, its dimer $(SO)_2$, and disulfur monoxide S_2O. There has been a great deal of confusion with respect to these oxides and what is now known to be S_2O was thought to be SO or $(SO)_2$. The most important of these oxides is sulfur monoxide, which is the crucial intermediate in all high temperature systems. SO is a highly reactive radical whose ground state is a triplet and electronically analogous to O_2. According to Cullis and Mulcahy (1972), its lifetime is seldom longer than a few milliseconds.

Spectroscopic studies have revealed other species in flames such as: CS, a singlet molecule analogous to CO and much more reactive; S_2, a triplet analogous to O_2 and the main constituent of sulfur vapor above 600°C; and the radical HS.

Johnson *et al.* (1970), calculated the equilibrium concentration of the various sulfur species for the equivalent of 1% SO_2 in propane–air flames. Their results, as a function of fuel–air ratio, are shown in Fig. 7. The dominance of SO_2 in the product composition for these equilibria calculations, even under deficient air conditions, should be noted. As reported earlier, practical systems reveal SO_3 concentrations (1–2%) which are higher than those depicted in Fig. 7.

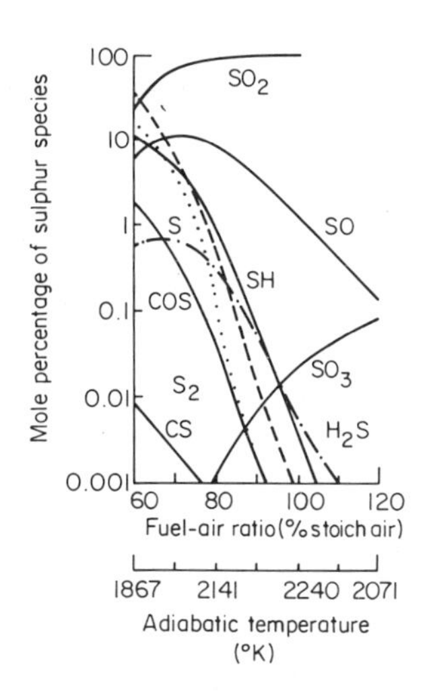

Fig. 7. Equilibrium distribution of sulfur-containing species in propane–air flames with unburnt gases initially containing 1% SO_2 (after Johnson *et al.*, 1951).

TABLE 2

Structure of gaseous sulfur compounds

Sulfur S_8	Rhombic	
Sulfur monoxide SO	S=O	
Sulfur dioxide SO_2 (OSO)		O=S=O (118°); $^{+}$S(=O)–O^{-}
Sulfur superoxide SO_2(SOO)	$^{+}$S≡O–O^{-}	
Sulfur trioxide SO_3		O=S(=O)=O (120°); O=S^{++}(O^{-})(O^{-})
Sulfur suboxide S_2O	S=S=O	
Carbonyl sulfide COS	S=C=O $^{-}$S–C≡O^{+} $^{+}$S≡C–O^{-}	
Carbon disulfide CS_2	S=C=S $^{-}$S–C≡S^{+}	
Organic thiols	R′–SH	
Organic sulfides	R′–S–R′	
Organic disulfides	R′–S–S–R′	

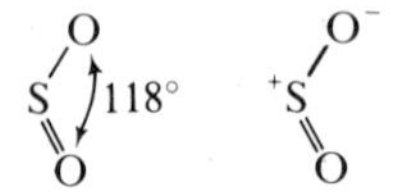

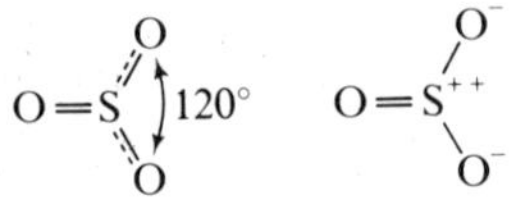

Insight into much that will be discussed and that which has been discussed can be obtained by the study of the structure of the various sulfur compounds given in Table 2.

2. Oxidative Mechanisms of Sulfur Fuels

Sulfur fuels characteristically burn with flames which are pale blue, sometimes very intensely so. This color comes about from emissions as a result of the reaction

$$O + SO \longrightarrow SO_2 + h\nu \tag{88}$$

and since it is found in all sulfur-fuel flames, this blue color serves to identify SO as an important reaction intermediate in all cases.

Most studies of sulfur–fuel oxidation have been performed using H_2S as the fuel. Consequently, the following material will concentrate on understanding the H_2S oxidation mechanism. Much of what is learned from these mechanisms can be applied to understanding the combustion of COS, CS_2, elemental and organically bound sulfur.

a. H_2S

Figure 8 is a general representation of the explosion limits of H_2S/O_2 mixtures. This three limit curve is very similar to that shown for H_2/O_2

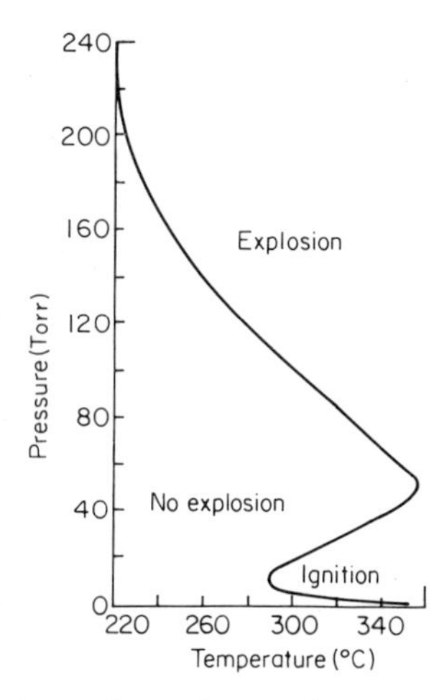

Fig. 8. Approximate explosion limits for stoichiometric mixtures of hydrogen sulfide and oxygen.

mixtures. However, there is an important difference in the character of the experimental data which determine the H_2S/O_2 limits. In the H_2S/O_2 peninsula and in the third limit region, explosion occurs after an induction period of several seconds.

The main reaction scheme for the low temperature oxidation of H_2S, although explicitly not known, would appear to be

$$\Delta H$$

$$H_2S + O_2 \longrightarrow SH + HO_2 \qquad 42 \text{ kcal/mole} \tag{89}$$

$$SH + O_2 \longrightarrow SO + OH \qquad -21 \text{ kcal/mole} \tag{90}$$

$$H_2S + SO \longrightarrow S_2O + H_2 \qquad -7 \text{ kcal/mole} \tag{91}$$

$$OH + H_2S \longrightarrow H_2O + HS \qquad -30 \text{ kcal/mole} \tag{92}$$

The addition of reaction (91) to this scheme is necessary because of the identification of S_2O in explosion limit studies. More importantly, Merryman and Levy (1971) in burner studies showed that S_2O occurs upstream from the peak of the SO concentration and that elemental sulfur was present still further upstream in the preignition zone.

The most probably system for the introduction of elemental sulfur is

$$SH + SH \longrightarrow H_2S + S \qquad -3 \text{ kcal/mole} \tag{93}$$

$$S + SH \longrightarrow S_2 + H \qquad -15 \text{ kcal/mole} \tag{94}$$

It does not seem feasible kinetically at the temperatures in the pre-ignition zone and at the overall pressures that the flame studies were carried out that the reaction

$$S_2 + O + M \longrightarrow S_2O + M \tag{95}$$

could account for the presence of S_2O. The disproportion of SO would have to give SO_2 as well as S_2O. Since SO_2 is not found in certain experiments where S_2O can be identified, then disproportion would not be feasible and reaction (91) appears to be the best candidate for the presence of S_2O.

Reaction (90) is the branching step in the mechanism. Davies and Walsh (1973) suggest that

$$SH + O_2 + M \longrightarrow HSO_2 + M \tag{96}$$

competes with reaction (90) and determines the second limit. Cullis and Mulcahy (1972) suggest the reaction

$$S_2O + O_2 \longrightarrow SO_2 + SO \tag{97}$$

as the degenerate branching step. The explicit mechanism for forming S_2O and its role in flame processes must be considered a great uncertainty.

At higher temperatures the reaction

$$O_2 + SO \longrightarrow O + SO_2 \tag{98}$$

becomes competitive with reaction (91) and introduces O radicals into the system. The presence of O radicals gives another branching reaction, namely,

$$O + H_2S \longrightarrow OH + SH \tag{99}$$

The branching is held in check by reaction (91) which removes SO, and the fast termolecular reaction

$$O + SO + M \longrightarrow SO_2 + M \tag{100}$$

which removes both O radicals and SO.

In shock tube studies, SO_2 is formed before OH radicals appear. To explain this result it has been postulated that the reaction

$$O + H_2S \longrightarrow SO + H_2 \qquad -53 \text{ kcal/mole} \tag{101}$$

is possible. This reaction and reaction (98) give the overall step

$$O_2 + H_2S \longrightarrow SO_2 + H_2 \tag{102}$$

Detailed sampling in flames by Sachjan *et al.* (1967), indicates that the H_2S is oxidized in a three-step process. During the first stage, most of the H_2S is consumed, and the products are mainly sulfur dioxide and water. In the second stage, the concentration of SO decreases, the concentration of OH reaches its maximum value, the SO_2 reaches its final concentration, and the concentration of the water begins to build as the hydrogen passes through a maximum.

The interpretation given to these results is that during the first stage the H_2S and O_2 are consumed mainly by reactions (101) and (98)

$$O + H_2S \longrightarrow SO + H_2 \tag{101}$$

$$SO + O_2 \longrightarrow SO_2 + O \tag{98}$$

with some degree of chain branching by reaction (99)

$$O + H_2S \longrightarrow OH + SH \tag{99}$$

In the second stage, reaction (98) predominates over reaction (101) because of the depletion of the H_2S and the OH concentration rises via reaction (99) and begins the oxidation of the hydrogen

$$O + H_2 \longrightarrow OH + H \tag{103}$$

Of course the complete flame mechanism must include

$$H + O_2 \longrightarrow OH + O \tag{104}$$

$$OH + H_2 \longrightarrow H_2O + H \tag{105}$$

Reactions (99) and (101) together with the fast reaction at the higher temperature

$$SO + OH \longrightarrow SO_2 + H \tag{106}$$

explain the known fact that H_2S inhibits the oxidation of hydrogen.

b. COS *and* CS_2

Even though there have been appreciably more studies of very flammable CS_2, COS is known to exist as an intermediate in CS_2 flames. Thus it appears logical to analyze the COS oxidation mechanisms first. Both substances show explosion limit curves which indicate that branched chain mechanisms exist. Most of the reaction studies used flash photolysis and very little information exists on what the chain initiating mechanism for thermal conditions would be.

COS flames exhibit two zones. In the first zone carbon monoxide and sulfur dioxide form, and in the second zone the carbon monoxide is converted into carbon dioxide. Since these flames are hydrogen free, it is not surprising to note that the CO conversion in the second zone is rapidly accelerated by adding a very small concentration of water to the system.

Photolysis initiates the reaction by creating sulfur atoms.

$$COS + h\nu \longrightarrow CO + S \quad (107)$$

The S atom then creates the chain branching step

$$S + O_2 \longrightarrow SO + O \quad -5 \text{ kcal/mole} \quad (108)$$

which is followed by

$$O + COS \longrightarrow CO + SO \quad -51 \text{ kcal/mole} \quad (109)$$

$$SO + O_2 \longrightarrow SO_2 + O \quad -3 \text{ kcal/mole} \quad (98)$$

At high temperatures the slow reaction

$$O + COS \longrightarrow CO_2 + S \quad -54 \text{ kcal/mole} \quad (110)$$

must also be considered.

The initiation step under purely thermally induced conditions such as those imposed by shocks has not been postulated, but is simply thought to be a reaction which produces O atoms. The high temperature mechanism would then be reactions (98), (108)–(110) with termination by the elimination of the O atoms.

For the explosive reaction of CS_2, Myerson *et al.* (1957) suggest

$$CS_2 + O_2 \longrightarrow CS + SOO \quad (111)$$

as the chain initiating step. Although the existence of the superoxide, SOO is not universally accepted, it is difficult to conceive a more logical thermal initiating step, particularly when the reaction can be induced in the 200–300°C range.

The introduction of the CS by reaction (111) starts the following chain scheme

$$CS + O_2 \longrightarrow CO + SO \quad -83 \text{ kcal/mole} \quad (112)$$

$$SO + O_2 \longrightarrow SO_2 + O \quad -13 \text{ kcal/mole} \quad (98)$$

$$O + CS_2 \longrightarrow CS + SO \quad -78 \text{ kcal/mole} \quad (113)$$

$$CS + O \longrightarrow CO + S \quad -77 \text{ kcal/mole} \quad (114)$$

$$S + O_2 \longrightarrow SO + O \quad -5 \text{ kcal/mole} \quad (108)$$

$$S + CS_2 \longrightarrow S_2 + CS \quad -6 \text{ kcal/mole} \quad (115)$$

$$O + CS_2 \longrightarrow COS + S \quad (116)$$

$$O + COS \longrightarrow CO + SO \quad (110)$$

$$O + S_2 \longrightarrow SO + S \quad (117)$$

The high flammability of CS_2 in comparison to COS is thought to be brought about by the greater availability of S atoms. At low temperatures, branching occurs in both systems via

$$S + O_2 \longrightarrow SO + O \quad (108)$$

Even greater branching occurs since in CS_2 one has

$$O + CS_2 \longrightarrow CS + SO \quad (113)$$

The comparable reaction for COS is reaction (110)

$$O + COS \longrightarrow CO + SO \quad (110)$$

which is not chain branching.

c. Elemental Sulfur

Elemental sulfur is found in all the flames of sulfur-bearing compounds discussed in the previous subsections. Generally this sulfur appears as atoms or the dimer S_2. When pure sulfur is vaporized at low temperatures, the vapor molecules are polymeric in nature and have the formula S_8. Vapor phase studies of pure sulfur oxidation around 100°C have shown that the oxidation reaction has the characteristics of a chain reaction. It is interesting to note that in the explosive studies the reaction must be stimulated by the introduction of O atoms (spark, ozone) in order for the explosion to proceed. Levy *et al.* (1970) report that Semenov suggests the following initiation and branching reactions:

$$S_8 \longrightarrow S_7 + S \quad (118)$$

$$S + O_2 \longrightarrow SO + O \quad (108)$$

$$S_8 + O \longrightarrow SO + S + S_6 \quad (119)$$

with the products produced by

$$SO + O \longrightarrow SO_2^* \longrightarrow SO_2 + h\nu \quad (88)$$

$$SO + O_2 \longrightarrow SO_2 + O \quad (98)$$

$$SO_2 + O_2 \longrightarrow SO_3 + O \quad (120)$$

$$SO_2 + O + M \longrightarrow SO_3 + M \quad (121)$$

A unique feature of oxidation of pure sulfur is that the percent of SO_3 formed is a very much larger (about 20%) fraction of the SO_2 than is generally found in the oxidation of sulfur compounds.

d. Organic Sulfur Compounds

It is more than likely that when sulfur is contained in a crude oil or in coal (other than the pyrites) it is organically bound in one of the three forms listed in Table 2—the thiols, sulfides, or disulfides. The combustion of these compounds is very much different than the other sulfur compounds studied in that a large portion of the fuel element is a pure hydrocarbon fragment. Thus in explosion or flame studies, the branched chain reactions which determine the overall consumption rate or flame speed would characteristically follow those chains described in hydrocarbon combustion and not the CS, SO, and S radical chains which dominate in H_2S, CO, COS, and S_8 combustion.

A major product in the combustion of all organic sulfur compounds is sulfur dioxide. Sulfur dioxide has a well-known inhibiting effect on hydrocarbon and hydrogen oxidation and, indeed, is responsible for a self-inhibition in the oxidation of organic sulfur compounds. This inhibition most likely arises from its role in the removal of H atoms by the termolecular reaction

$$H + SO_2 + M \longrightarrow HSO_2 + M \qquad (122)$$

HSO_2 is a known radical which has been found in H_2–O_2–SO_2 systems and is sufficiently inert to be destroyed without reforming any active chain carrier.

In the lean oxidation of the thiols, even at temperatures around 300°C, all the sulfur is converted to SO_2. At lower temperatures and under rich conditions, disulfides form and other products such as aldehydes and methanol are found. The presence of the disulfides suggests a chain-initiating step very much similar to low temperature hydrocarbon oxidation

$$RSH + O_2 \longrightarrow RS + HO_2 \qquad (123)$$

Cullis and Mulcahy report this step to be followed by

$$RS + O_2 \longrightarrow R + SO_2 \qquad (124)$$

to form the hydrocarbon radical and sulfur dioxide. One must question whether the SO_2 in reaction (124) is sulfur dioxide or not. If the O_2 strips the sulfur from the RS radical, then it is more likely the SO_2 is the sulfur superoxide, which would decompose or react to form SO. The SO is then oxidized to sulfur dioxide as described previously. The organic radical is oxidized, as discussed in Chapter 3. The radicals formed in the subsequent oxidation, of course, attack the original fuel to give the RS radical, and the initiating step is no longer needed. The disulfide most likely forms from two RS radicals reacting

$$RS + RS \longrightarrow RSSR \qquad (125)$$

The principal products in the oxidation of the sulfides are sulfur dioxide and aldehydes. The low temperature-initiating step is similar to reaction (123) except the hydrogen abstraction is known to be from the carbon atom next to the sulfur atom, i.e.,

$$RCH_2SCH_2R + O_2 \longrightarrow RCH_2S{-}CHR + HO_2 \tag{126}$$

The radical formed in reaction (126) then decomposes to form an alkyl radical and a thioaldehyde molecule, i.e.,

$$RCH_2S{-}CHR \longrightarrow RCH_2 + RCHS \tag{127}$$

Both products in reaction (127) are then oxidized. The thioaldehydes give SO_2 and aldehydes as products. How the aldehydes form is not exactly clear, but they are subsequently further oxidized.

The oxidation of the disulfides follow a similar route to the sulfide with an initiating step

$$RCH_2SSCH_2R + O_2 \longrightarrow RCH_2S{-}S{-}CHR + HO_2 \tag{128}$$

followed by radical decomposition

$$RCH_2S{-}SCH{-}R \longrightarrow RCH_2S + RCHS \tag{129}$$

The thiol is then formed by hydrogen abstraction

$$RCH_2S + RH \longrightarrow RCH_2SH + R \tag{130}$$

and the oxidation proceeds as described previously.

e. Sulfur Trioxide and Sulfates

Earlier it was pointed out that the concentration of sulfur trioxide found in the combustion gases of flames, though small, is greater than that which would have been expected from equilibrium calculations. Indeed this same phenomenon exists in large combustors, such as furnaces, in which there is a sulfur component in the fuel used. The equilibrium represented by Eq. (87)

$$SO_2 + \tfrac{1}{2}O_2 \rightleftharpoons SO_3$$

is shifted strongly to the left at high temperatures and one would expect very little SO_3 in a real combustion environment. It is very apparent, then, that the combustion chemistry involved in oxidizing sulfur dioxide to the trioxide is such that equilibrium cannot be obtained.

Truly the most interesting finding is that the superequilibrium concentrations of SO_3 are very sensitive to the original oxygen concentration. Under fuel-rich conditions approaching even stoichiometric conditions, practically no SO_3 is found. In proceeding from stoichiometric to 1% excess air, a sharp increase in the conversion of SO_2 to SO_3 is found. Further

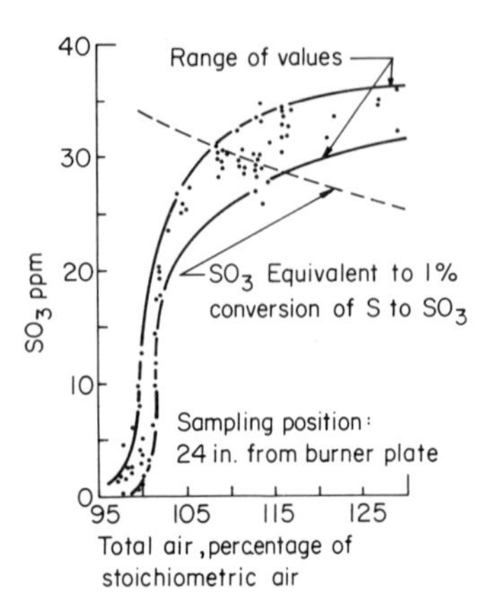

Fig. 9. Effect of excess air on the formation of sulfur trioxide in a hydrocarbon flame (after Barrett *et al.*, 1966).

addition of air only causes a slight increase, however, the effect of the excess nitrogen in reducing the temperature could be a moderating factor in the rate of increase. Figure 9 taken from the work of Barrett *et al.* (1966) on hydrocarbon flames characterizes the results generally found not only in flame studies but also with furnaces. Such results strongly indicate that the SO_2 is converted into SO_3 in a termolecular reaction with oxygen atoms

$$O + SO_2 + M \longrightarrow SO_3 + M \tag{131}$$

It is important to note that the superequilibrium results are obtained with either sulfur fuels, small concentration of sulfur fuels added to hydrocarbons, SO_2 added to hydrocarbon, etc. Further confirmation supporting reaction (131) as the conversion route comes from the observation that in carbon monoxide flames the amount of SO_3 produced was substantially higher than in all other cases. It is well known that since O atom cannot attack CO directly, the SO_3 concentration is much higher in CO flames than any other flames. The fact that in all cases the SO_3 concentration also increases with pressure supports a termolecular route such as reaction (131).

It is well known that the thermal dissociation of SO_3 is slow and that the concentration of SO_3 is then frozen within its stay time in flames and furnaces. The thermal dissociation rates are known, but one can also calculate the superequilibrium concentration of oxygen atom in flames. If so, then the SO_3 concentration should correspond to the equilibrium concentration given by

$$O + SO_2 + M \rightleftharpoons SO_3 + M \tag{132}$$

in which the oxygen atom superequilibrium concentration is used. However, the SO_3 concentrations are never this high, thus one must conclude that some SO_3 is being reduced by routes other than thermal decomposition. The two most likely routes are by oxygen and hydrogen atom attack on the sulfur trioxide via

$$O + SO_3 \longrightarrow O_2 + SO_2 \qquad -37 \text{ kcal/mole} \tag{133}$$

$$H + SO_3 \longrightarrow OH + SO_2 \qquad -19 \text{ kcal/mole} \tag{134}$$

Evidence supports this contention, and it would appear that reaction (133) would be more important than reaction (134) in controlling the SO_3 concentration with reaction (132). Further, it is important to note that reactions (132) and (133) are effective means of reducing O radical concentration. Since reaction (122) has been shown to be an effective means of reducing H radical concentrations, one can draw the general conclusion that sulfur compounds reduce the extent of superequilibrium concentration of the characteristic chain carrying radicals which exist in hydrocarbon flames.

In dealing with furnaces using residual oils, one must recognize heterogeneous catalysis as a possible route for the conversion of SO_2 to SO_3. Sulfur dioxide and molecular oxygen will react catalytically on steel surfaces and vanadium pentoxide (deposited from vanadium compounds in the fuel) and at lower temperatures where the equilibrium represented by reaction (82) favors the formation of SO_3.

If indeed SO_2 and SO_3 are effective in reducing the superequilibrium concentration of radicals in flames, then it is apparent that sulfur compounds should play a role in NO formation from atmospheric nitrogen in flame systems. Since no matter what type of sulfur compounds is added to combustion systems SO_2 and SO_3 form, these species reduce the oxygen atom concentration and thus should reduce NO formation. Wendt and Ekmann (1975) have recently reported flame data which appear to substantiate this conclusion.

In examining reactions (132) and (133), one realizes that SO_2 plays a role in catalyzing the recombination of oxygen atoms. Indeed this homogeneous catalytic recombination of oxygen atoms causes the decrease in the superequilibrium concentration of the oxygen atoms. SO_2 also plays a role in the recombination of hydrogen radicals through the route

$$H + SO_2 + M \longrightarrow HSO_2 + M \tag{122}$$

$$H + HSO_2 \longrightarrow H_2 + SO_2 \tag{135}$$

and the recombination of hydrogen and hydroxyl radicals through the route of reaction (122) and

$$OH + HSO_2 \longrightarrow H_2O + SO_2 \tag{136}$$

D. PARTICULATE FORMATION

In earlier sections of this chapter, the role that particulates would play in a given environmental situation was identified. This section will be devoted exclusively to those particulates which form in combustion processes and which primarily have carbon as the main constituent. Those carbon particulates which form from gas phase processes are generally referred to as soot and those which develop from pyrolysis of the liquid hydrocarbon fuels are generally referred to as coke or cenospheres.

Although there are various restrictions to carbon particulate output from different types of power plants, these particles can play both a beneficial and detrimental role in an overall industrial process. The presence of particulates in gas turbines can severely affect the lifetime of the turbine blades. However, in many industrial furnaces the presence of carbon particulates can appreciably increase the rate of heat transfer from the combustion gases by increasing the radiative component.

This last point is worth considering in somewhat more detail. Most flames are luminous and this luminosity arises from the fact that there are carbon particles present and they radiate strongly at the high combustion gas temperatures. The condensed phase particulates have a very high emissivity compared to the gaseous products formed and thus appreciably increase the radiant heat transfer. In fact some systems can approach blackbody conditions. Most flames characteristically appear yellow when there is particulate formation. The observance of the yellow radiation is the indication of fuel-rich operation and the yellow color is due to the fact that Planck's radiation curve peaks in the yellow region of the electromagnetic spectrum for the temperatures normally found in combustion systems. Thus in certain industrial furnaces in which the rate of heat transfer from the combustion gases to some surface, such as a melt, is important, it is beneficial to operate the system under fuel-rich conditions to assure the presence of the carbon particles. In many instances these carbon particles could be burned with extra air in later stages in order to meet environmental restrictions. Indeed some luminous flames are nonsooting because the very small particles formed are oxidized before leaving the flame zone.

It has been well known that the extent of carbon particulate formation is related to the type of flame existing in a given process. Diesel exhausts are known to be more smokey than those of carbureted spark ignition engines. Diffusion flames prevail in fuel injected diesel engines, whereas carbureted spark ignition engines entail the combustion of nearly homogeneous premixed fuel–air systems.

Until recently most discussions of carbon particulate formation dealt primarily with the conditions under which such particulates form. More recently, however, fundamental research has been elucidating many of the individual processes. In particular, one should refer to the book by Weinberg and Lawton (1969) and the papers by Homann and Wagner referenced in their article in the Eleventh Combustion Symposium (1967). Two excellent reviews which have been drawn upon heavily in writing this section are by Palmer and Cullis (1965) and by Schalla and Hibbard in NACA Report 1300 (1959). Bradley (1969) has a brief section on the subject as well. Most of the following discussion will deal with carbon formation involving a gaseous system, however, at the end of the chapter a short discussion on the cokelike formation from liquid fuels will be given.

1. Characteristics of Soot

The characteristics of soot are well described in the article by Palmer and Cullis (1965) who give detailed references on the subject matter. Aspects of their review are worth summarizing directly.

Investigators have used the words "carbon" and "soot" to describe a wide variety of solid materials, many of which contain appreciable amounts of hydrogen and other elements and compounds which may have been present in the original hydrocarbon fuel. The properties of the solids change markedly with the conditions of formation and indeed several quite well-defined varieties of solid carbon may be distinguished. One of the most obvious and important differences depends on whether the carbon is formed by a homogeneous vapor-phase reaction or whether the carbon is deposited on a solid surface which may be present in or near the reaction zone.

The existence of two distinct types of carbon which arise due to the presence of a surface or not was first noted on studies of acetylene decomposition flames. It was further noted that the presence of small amounts of oxygen or water vapor in the decomposing gases largely suppresses the formation of the gas-phase carbon.

Investigations have been made both on the properties of the carbon formed and the extent of the carbon formation for various experimental conditions and for both diffusion and premixed flames. In general, however, the properties of the carbon particulates formed in flames are remarkably little affected by the type of the flame, the nature of the fuel being burned and the other conditions under which they may have been produced. The extent of carbon particulates does vary appreciably with flame type. Diffusion flames invariably give more carbon than premixed flames. Theories must be able to explain these experimental findings.

Palmer and Cullis (1965) report the detailed characteristics as follows.

"The carbon formed in flames generally contains at least 1% by weight of hydrogen. On an atomic basis this represents quite a considerable proportion of this element and corresponds approximately to an empirical formula of C_8H. When examined under the electron microscope, the deposited carbon appears to consist of a number of roughly spherical particles, strung together rather like pearls on a necklace. The diameters of these particles vary from 100 to 2000 Å and most commonly lie between 100 and 500 Å. The smallest particles are found in luminous but nonsooting flames, while the largest are obtained in heavily sooting flames. x-ray diffraction shows that each particle is made up of a large number ($\sim 10^4$) of crystallites. Each crystallite is shown by electron diffraction to consist of 5–10 sheets of carbon atoms (of the basic type existing in ideal graphite), each containing about 100 carbon atoms and thus having length and breadth of the order of 20–30 Å; but the layer planes, although parallel to one another and at the same distance apart, have a turbostratic structure, i.e., they are randomly stacked in relation to one another, with the result that the interlayer spacing (3.44 Å) is considerably greater than in ideal graphite (3.35 Å). It may readily be calculated on this picture of dispersed carbon deposits that an "average" spherical particle contains from 10^5 to 10^6 carbon atoms. The fundamental problem for elucidation is the mechanism whereby simple fuel molecules containing only a few carbon atoms are converted so rapidly into these high aggregates.

"The simplest test for carbon formation in flames is the luminosity, which was shown long ago by Davy to be due to the presence of carbon particles. Other solid particles appear at the top of a flame as soot; but even where a luminous flame does not actually give soot as such, it is frequently possible to obtain a carbon deposit by passing a cold probe through the luminous zone. Luminosity has been used as a quantitative measure of carbon formation, but a more convenient and widely used criterion is the flame height at which luminosity or soot formation just becomes apparent. Thus, although small diffusion flames on circular burners do not show carbon formation, an increase in the fuel flow causes the flame to develop a luminous tip at a fairly definite height. The luminous region first appears not at the tip, but at a point within the otherwise blue flame area. As the fuel flow is further increased, the luminous part gets longer until it extends beyond the boundary of the blue flames and finally sooting starts at the top. The flame height at which soot formation starts is generally the most convenient (inverse) measure of the tendency to form carbon since, unlike the flame height at which luminosity first becomes apparent (which is usually too small for accurate measurement), it is farily large and easy to measure."

2. Mechanisms of Soot Formation in Flames

The fact that certain flames are luminous, but not soot forming, is clear evidence that soot formation depends on the competitive rates of reactions forming soot and those which cause its oxidation. One can consider that there are three stages to the soot-forming reaction. The first stage is a most complex one called nucleation and one in which gas-phase reactions occur and lead to condense-phase solid nuclei. This stage is followed by heterogeneous reactions which occur on the surface of these nuclei. These surfaces are very active in catalyzing such reactions because of the presence of free valences. The final stage is one of agglomeration or coagulation and is one which is slow compared to the other two. Only a few of the intermediate species which undergo heterogeneous reaction in the second stage are involved in the original nucleation process.

The nucleation stage is quite apparently the most important one. Although it was inferred above that the nuclei form from gas-phase reactions, Weinberg (1962) in his work on the effect of electric fields on flames has shown the importance of positive ions in the process. It appears that carbon particles will grow on positive ions formed in the reaction zone of the flame just as they do on normal nuclei. Weinberg's work shows that carbon particles acquire a positive charge when they are very small. As a result they can be manipulated by an electric field and their size controlled by adjustment of their residence time in what can be called the pyrolysis zone of the flame.

The present conceptual process about carbon nuclei formation entails the sequential, but partially concurrent processes of the pyrolysis of the fuel molecules to lower molecular weight fragments in the same context as was described in Chapter 3 in which the oxidation of higher-order hydrocarbon was discussed, the polymerization of these fragments and continued dehydrogenation. It is important to note again that it was found that the aliphatic hydrocarbons, even in the presence of fairly substantial amounts of oxygen, break down to give the olefins. Polymerization requires such unsaturated compounds. In other carbon processes, in order to account for the relatively high rate of formation of carbon, one must conclude that the reactions involved must be of the radical–molecule type.

Homann and Wagner's work with premixed flames, primarily low pressure acetylene and benzene flames, show that in flames of both aliphatic and aromatic fuels, there are mainly three groups of hydrocarbons that can be considered to take an active role in the process of carbon formation:

(1) acetylene and polyacetylenes (mass range 26 to 146);
(2) polycyclic aromatic hydrocarbons (mass range 79 to $\sim$300);

(3) reactive polycyclic hydrocarbons, probably with side chains which contain more hydrogen than aromatics.

The critical intermediates in the process must meet the rather unique requirement of being highly stable at flame temperatures and having a high reactivity toward polymerization.

The carbon formation scheme proposed for acetylene is shown in Table 3.

TABLE 3

Carbon formation route from acetylene flame

Acetylene	Radical reactions → with C_2H and C_2H_3	Polyacetylenes → polyacetylene radical
addition → of a radical	Branched radical	Addition of C_2H_2 and → polyacetylene, cyclization
→	Polycyclic aromatic hydrocarbons ↑ Reactive partly cyclic hydrocarbons, hydrogen-rich (group 3)	Further addition → of polyacetylene
→	Small soot particles (active)	Addition of small soot → particles and polyacetylenes process, unactivative
→	Polycyclic aromatic by surface reactions ↑ Large soot particles (inactive 250 Å)	→
→	Agglomoration large soot particles to aggregates, slow growth of carbon amount by heterogeneous decomposition of C_2H_2 and polyacetylenes	

It is inferred, following the scheme shown in this table, that aliphatic fuels are converted to acetylene first, and then polymerization to polyacetylene and polyacetylene radicals takes place. As the chain length increases, cyclization increases and becomes more important. Provided that the radical character is retained in the reacting sequence, reaction with polyacetylene continues. It is not surprising then that with aromatic fuels, in which the polycyclic aromatics hydrocarbons can occur much earlier in the flame where radical species are more abundant, that the tendency to carbon formation is greater. A free radical-benzene carbon forming scheme given by Gordon *et al.* (1959) is shown in Table 4.

TABLE 4

Carbon formation from benzene flames

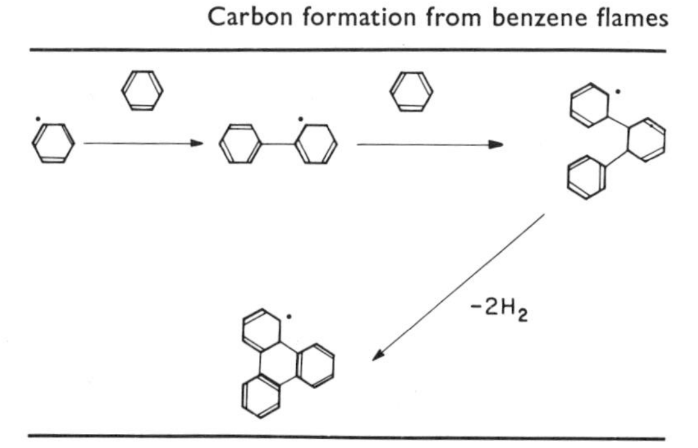

Although based on experiments with premixed flames, the previous discussion holds for diffusion flames as well. In premixed flames one would expect that carbon formation would take place when the carbon to oxygen ratio is insufficient for conversion of the carbon in the fuel to carbon monoxide. In fact, carbon formation is found in mixtures in which the oxygen is more than sufficient for this conversion. More recently, Glassman *et al.* (1975) has shown that in the oxidation of aliphatic hydrocarbons not only is the first step the conversion to the olefin, but in a relatively fuel-rich system acetylene can be identified later in the process.

Many have attributed this high carbon-to-oxygen ratio effect to the Boudouard reaction which is the disproportionation of carbon monoxide. This reaction may be written as

$$2CO \rightleftharpoons CO_2 + C_S$$

The equilibrium calculation of the products of ethylene oxide decomposition in a monopropellant would indicate extensive carbon formation. However such is not found, probably due to the short stay time of the reacting species in the rocket motor. The equilibrium in Boudouard reaction shifts to the right at low temperatures and well over to the left about 1000°K. The rate of reaction at low temperatures is very slow, and indeed every condensation reaction has a relatively slow rate, so that carbon formation by this step is not noted in processes which have short stay times such as those which exist in rocket motors. There are conditions under which it could play a role such as in a boundary layer at temperatures slightly lower than 1000°K.

3. The Influence of Physical and Chemical Parameters on Soot Formation

The previous discussion makes it apparent that the molecular structure of the fuel is the dominant factor in soot formation. But, recall, it was stated earlier that the presence of soot depends upon the competitive reaction rates of the nucleation step and those which oxidize the intermediates. So it should not be surprising that there will be different ordering of the molecular structure whether one considers diffusion flames or premixed flames. Across any plane through a diffusion flame, there is a wide variation of fuel–oxidizer ratio from very fuel-rich to very fuel-lean. Thus in diffusion flames there is always a zone very close to the flame that is at high temperature and which has a very high carbon to oxygen ratio. This characteristic of diffusion flames is the reason that they always have some luminosity and form soot relatively easily. A premixed flame must be very fuel-rich to become luminous and soot. In diffusion flames soot formation always appears at the tip first. This observation would seem to indicate that there must be accumulation of the reactive intermediates, which are initiated at the burner exit, and flow towards the flame tip.

In diffusion-type flames carbon formation decreases in the order

naphthalenes > benzenes > acetylenes > di-olefins > mono-olefins > paraffins

or in the more general order,

aromatics > alkynes > alkenes > alkanes

which is perfectly consistent with the mechanism of formation described.

That the decrease of unsaturation determines the tendency towards carbon formation is further confirmed by the finding that for a given organic series increased molecular weight decreases carbon formation. As one might intuitively expect, it also has been found that branched-chain compounds have a greater tendency to liberate carbon than do straight-chain compounds. With alcohols and paraffins the tendency to smoke (other than methanol which does not smoke) increases with the number of carbon atoms. These effects are graphically depicted in Fig. 10. Heuristically, one could conclude that the C/H ratio is one of the principal factors controlling the tendency to soot, but that the compactness of the molecule can play an important part. These conclusions are essentially consistent with the general mechanisms given in the preceding subsections.

In premixed flames the smoke tendency order with respect to molecular structure is different and found to be

naphthalenes > benzenes > alcohols > paraffins > olefins > aldehydes > acetylenes

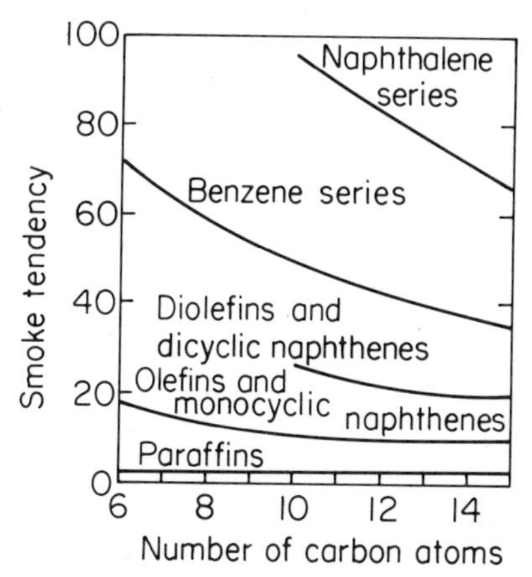

Fig. 10. Tendency of various fuels to smoke as measured by the reciprocal height to soot formation in a diffusion flame (after Clarke et *al.*, 1946).

Clearly the large change in order can be almost precisely explained by considering the reactivity of the fuel with oxygen. Recall rapid oxidation of the fuel and fuel intermediates can remove the polymerization precursors. Acetylenes, aldehydes, and olefins are very reactive at high temperatures, whereas the aromatic compounds still are not.

With regard to this last point about activity, it is most interesting to note that soot formation is drastically reduced in turbulent diffusion flames. High intensity turbulence can undoubtedly rapidly mix the fuel and oxidizer in diffusion flames and increase the oxidation of the soot precursors. Similarly the tendency to soot decreases in turbulent premixed flames. These concepts are entirely consistent with discussion of turbulent flames given in Section 4.C.

In all type flames the lower the pressure the lower the tendency to soot. The pressure sensitivity is consistent with lowering the rates of all the reactions in the nucleation process. The effect of inert diluents is similar in that they effectively lower the partial pressure of the precursor reactants.

The effects of possible reactive additives are most interesting. In diffusion flames, sulfur trioxide, gaseous hydrogen and nickel and the alkaline earth salts tend to suppress carbon formation somewhat. Indeed barium salts were added to a commercial gasoline as a soot suppressant. Interestingly in a premixed coal gas flame the addition of sulfur trioxide can markedly increase the luminosity. Many of these effects can be explained by considering some recent work by Feugier (1976) on the action of metallic additives on soot formation in rich, premixed, ethene–oxygen–nitrogen flames.

Feugier found that the quantity of soot from mildly sooting flames increased greatly with the addition of sodium, potassium, and cesium salts.

The lower the ionization potential of the metal the greater was the tendency to soot. These observations appear consistent with Weinberg's suggestion that positive ions are excellent sites for nuclei growth.

In the case of lithium, barium, strontium, and calcium salts, he found a strong inhibiting effect on soot formation. In all cases the greater the percentage of metal additive the greater the reduction in soot formation. One exception is noted and that is for barium, where a minimum reduction is reached and then the soot tendency increases. In one case for relatively large amounts of barium salt, there was augmentation over the initial amount obtained without any additive. The inhibition by the lithium and alkaline–earth salts can be explained by considering the competition between precursor oxidation and the nucleation reactions.

It is well known that lithium readily forms the hydroxide according to the reaction:

$$\mathrm{Li + H_2O \longrightarrow LiOH + H}$$

Similarly the alkaline–earths form their hydroxides from the sequence

$$\mathrm{M + H_2O \longrightarrow MOH + H}$$

$$\mathrm{MOH + H_2O \longrightarrow M(OH)_2 + H}$$

in which M represents the alkaline–earth metal. It is then postulated that the H radicals formed in these reactions attack a water molecule to produce the hydroxyl radical:

$$\mathrm{H + H_2O \longrightarrow OH + H_2}$$

The hydroxyl could not form from the attack of the H radicals on oxygen molecules since all systems are rich and deficient in O_2. It is then postulated that the hydroxyl radicals readily oxidize the soot precursors to prevent carbon nucleation.

The different effects of SO_3 on diffusion and premixed flames can be explained in the context of the last paragraph and the discussion of the sulfur oxide reactions in the preceding section. In diffusion flames SO_3 tends to suppress carbon formation slightly. The precursors to the carbon nuclei undoubtedly form on the fuel side of a diffusion flame. There must be some H radicals present which react with the SO_3

$$\mathrm{H + SO_3 \longrightarrow HO + SO_2}$$

to give OH radicals. As discussed previously, these hydroxyl radicals oxidize the precursors. In premixed flames the SO_3 must dissociate into SO_2 which removes H radicals by reaction (117):

$$\mathrm{H + SO_2 + M \longrightarrow HSO_2 + M} \tag{117}$$

and thus reduces the concentration of hydroxyl radicals formed in the rich premixed system.

4. Particulates from Liquid Hydrocarbon Pyrolysis

The effluent from many liquid-fueled combustion systems contains, in many instances, not only soot formed from the gas-phase processes described in the previous subsections, but also carbon particulates that are formed from the liquid hydrocarbon itself. If a liquid spray is allowed to impinge directly on a very hot wall before it is appreciably oxidized then the hydrocarbon is known to undergo liquid-phase cracking, subsequent pyrolysis, and finally coking. The resulting carbon particulate called petroleum coke is a harder material than the soot formed from gas–phase processes. Experiments burning heavy multicomponent residual fuels have shown that in the latter stages of burning of these fuel droplets a coke will form. These coke particles have been referred to as cenospheres and can be difficult to burn. They are very much like, if not the same as, petroleum coke. Unfortunately, as compared to that which has been performed in determining the gas-phase mechanisms of soot formation, much less work has been done in this area. Early work (Berry and Edgeworth-Johnson, 1944) shows the following possible sequence of reactions to form this petroleum coke:

paraffins ⟶ olefins ⟶ aromatics with side chain ⟶
condensed ring systems ⟶ asphalt ⟶ pitch ⟶
semipitch ⟶ asphaltic coke ⟶ carboid coke.

As one would readily realize, the fuels containing aromatics form petroleum coke more readily than other fuels. In regenerative cooled liquid propellant rockets, one must remove the aromatics from the fuel or coke forms in the cooling passage at the wall near the injector. This thermal insulating coke formation can cause burn-out of the rocket wall. Indeed, rocket propulsion fuels differ from the jet propulsion fuels mainly in aromatic content.

E. STRATOSPHERIC OZONE

The nature of the ozone balance in the stratosphere is determined through complex interactions of solar radiation, meteorological movements within the stratosphere, transport to and from the troposphere, and the concentration of species which are based on elements other than oxygen and which arrive into the stratospheric by natural or artificial means (such as flight of aircraft).

It is not difficult to perceive that ozone initially forms from the oxygen present in the air. Chapman (1930) initially developed the photochemical model of stratospheric ozone and suggested that the ozone mechanism depended on two photochemical and two chemical reactions:

$$O_2 + h\nu \longrightarrow O + O \tag{137}$$

$$O + O_2 + M \longrightarrow O_3 + M \tag{138}$$

$$O_3 + h\nu \longrightarrow O_2 + O \tag{139}$$

$$O + O_3 \longrightarrow O_2 + O_2 \tag{140}$$

Reactions (137) and (138) are the reactions by which the ozone is created. Reactions (139) and (140) establish the balance which is the ozone concentration in the troposphere. If one adds reactions (139) and (140), the overall rate of destruction of the ozone is obtained, namely,

$$O_3 + h\nu \longrightarrow O_2 + O \tag{139}$$

$$O + O_3 \longrightarrow O_2 + O_2 \tag{140}$$

$$\text{net} \quad 2O_3 + h\nu \longrightarrow 3O_2 \tag{141}$$

The rates of reactions (137)–(140) vary with the altitude. The rate constants of reactions (137) and (139) are determined by the solar flux at a given altitude and rate constants of the other reactions by the temperature at that altitude. Precise solar data obtained from rocket experiments and better kinetic data for reactions (138)–(140) coupled with recent meteorological analysis showed that the Chapman model was in serious error. The concentrations predicted by the model were essentially too high. Something else was affecting the ozone.

1. The HO_x Catalytic Cycle

Hunt (1965, 1966) suggested that perhaps excited electronic states of O atoms and O_2 could account for the discrepancy between the Chapman model and the measured meteorological ozone distributions. But he showed that reactions based on these excited species were too slow to account for the differences sought. Realizing that water could enter the stratosphere, Hunt considered the reactions of free radicals (H, HO, HOO) derived from water. Consistent with the shorthand developed for the oxides of nitrogen, these radicals are specified by the chemical formula HO_x. The mechanism that Hunt postulated was predicated on the formation of hydroxyl radials. The photolysis of ozone by ultraviolet radiation below 310 nm produces excited singlet oxygen atoms which react rapidly with water to form hydroxyl radicals:

$$O_3 + h \longrightarrow O_2 + O(1D) \tag{142}$$

$$O(1D) + H_2O \longrightarrow 2OH \tag{143}$$

Only an excited singlet oxygen atom could react with water at stratospheric temperatures to form hydroxyl radicals.

At these temperatures, singlet oxygen atoms could also react with hydrogen or methane to form OH. The OH reacts with O_3 to produce hydroperoxy radicals HO_2. Both HO and HO_2 destroy ozone by an indirect reaction which sometimes involves O atoms:

$$HO + O_3 \longrightarrow HO_2 + O_2 \tag{144}$$

$$HO_2 + O_3 \longrightarrow HO + O_2 + O_2 \tag{145}$$

$$HO + O \longrightarrow H + O_2 \tag{146}$$

$$HO_2 + O \longrightarrow HO + O_2 \tag{147}$$

$$H + O_3 \longrightarrow HO + O_2 \tag{148}$$

There are numerous reactions of HO_2 radicals possible in the stratosphere. The essential reactions for the discussion of the ozone balance are

$$HO + O_3 \longrightarrow HO_2 + O_2 \tag{144}$$

$$HO_2 + O_3 \longrightarrow HO + O_2 + O_2 \tag{145}$$

$$\text{net} \quad 2O_3 \longrightarrow 3O_2$$

The reaction sequence (144)–(145) is a catalytic chain for ozone destruction and contributes to the net destruction. However, even given the uncertainty possible in the rates of these reactions and the uncertainty of the air motions, this system could not explain the imbalance in the ozone throughout the stratosphere.

2. The NO_x Catalytic Cycle

In the late 1960s, direct observations of substantial amounts (3 ppb) of nitric acid vapor in the stratosphere were reported. Crutzen (1970) reasoned that if HNO_3 vapor is present in the stratosphere it could be broken down to a degree to the active oxides of nitrogen (NO_x–NO and NO_2) and that these oxides could form a catalytic cycle for the destruction of the ozone. Johnston (1971) first realized that if this were so, then supersonic aircraft flying in the stratosphere could wrought serious harm to the ozone balance in the stratosphere. Much of what appears in this section is drawn from an excellent review by Johnston and Whitten (1973). The most important of the possible NO_x reactions in the atmosphere for purposes here are:

$$NO + O_3 \longrightarrow NO_2 + O_2 \tag{146}$$

$$NO_2 + O \longrightarrow NO + O_2 \tag{147}$$

$$NO_2 + h\nu \longrightarrow NO + O \tag{148}$$

whose rate constants are now quite well known. The reactions combine in two different competing cycles. The first is catalytic destructive

$$NO + O_3 \longrightarrow NO_2 + O_2 \tag{146}$$

$$NO_2 + O \longrightarrow NO + O_2 \tag{147}$$

$$\text{net} \quad O_3 + O \longrightarrow O_2 + O_2$$

and the second parallel one is essentially a "do-nothing" one:

$$NO + O_3 \longrightarrow NO_2 + O_2 \tag{146}$$

$$NO_2 + h\nu \longrightarrow NO + O \tag{148}$$

$$O + O_2 + M \longrightarrow O_3 + M \tag{138}$$

net no chemical change

The rate of destruction of ozone with the oxides of nitrogen relative to the rate in "pure air" (Chapman model) is defined as the catalytic ratio, which may be expressed either in terms of the variables NO_2 and O_3 or NO and O. These ratio expressions are

$$\beta = \frac{\text{rate of ozone destruction with } NO_x}{\text{rate of ozone destruction in pure air}} \tag{149}$$

$$= 1 + \frac{k_{149}(NO_2)}{k_{140}(O_3)} \tag{150}$$

$$= 1 + \frac{(k_{146}k_{147}/k_{140}j_{148})(NO)}{1 + k_{147}(O)/j_{148}} \tag{151}$$

As throughout this book, the k's are the specific rate constants of the chemical reactions. The j's are the specific rate constants of the photochemical reactions.

At low elevations where the oxygen atom concentration is low and the NO_2 cycle slow, another catalytic cycle derived from the oxides of nitrogen may be important:

$$NO_2 + O_3 \longrightarrow NO_3 + O_2 \tag{152}$$

$$NO_3 + h\nu \text{ (visible, day)} \quad NO + O_2 \tag{153}$$

$$NO + O_3 \longrightarrow NO_2 + O_2 \tag{146}$$

$$\text{net} \quad 2O_3 + h\nu \longrightarrow 3O_2 \text{ (day)}$$

The radiation involved here is red light, which is abundant at all elevations. Reaction (152) permits another route at night (including the polar night) which converts a few percent of NO_2 to N_2O_5:

$$NO_2 + O_3 \longrightarrow NO_3 + O_2 \qquad (152)$$

$$NO_2 + NO_3 + M \longrightarrow N_2O_5 + M \quad \text{(night)} \qquad (154)$$

The rate of reaction (152) is only known accurately at room temperature, and extrapolation to stratospheric temperature is uncertain; nevertheless, the extrapolated values indicate the NO_3 catalytic cycle (reactions (152) and (153)) destroys ozone faster than the NO_2 cycle below 22 km and in the region where the temperature is at least 220°K.

The nitric acid cycle is established by the reactions

$$HO + NO_2 + M \longrightarrow HNO_3 + M \qquad (155)$$

$$HNO_3 + h\nu \longrightarrow OH + NO_2 \qquad (156)$$

$$HO + HNO_3 \longrightarrow H_2O + NO_3 \qquad (157)$$

The steady state concentration of nitrogen dioxide to nitric acid can be readily found to be

$$[(NO_2)/(HNO_3)]_{SS} = [k_{157}/k_{155}] + [j_{156}/k_{155}(OH)] \qquad (158)$$

For the best data available for the hydroxyl radical concentration and the rate constants, the ratio has the values

$$0.1 \text{ at } 15 \text{ km}, \quad 1 \text{ at } 25 \text{ km}, \quad \gg 1 \text{ at } 35 \text{ km}$$

Thus it can be seen that nitric acid is a significant reservoir or sink for the oxides of nitrogen. In the lowest stratosphere, the nitric acid predominates over the NO_2 and there is a major loss of NO_x from the stratosphere by diffusion of the acid into the troposphere where it is rained out.

By using the HO_x and NO_x cycles discussed and by assuming NO_x concentration 4.2×10^9 molecules/cm^3 distributed uniformly through the stratosphere, Johnston and Whitten (1973) were able to make the most reasonable prediction of the ozone balance in the stratosphere. The only measurements of the concentration of NO_x in the stratosphere show a range of $2\text{–}8 \times 10^9$ molecules/cm^3.

It is possible to similarly estimate the effect of the various cycles for ozone destruction. The results can be summarized as follows: between 15 and 20 km, the NO_3 catalytic cycle dominates; between 20 and 40 km, the NO_2 cycle; between 40 and 45 km the NO_2, HO_x, and O_x mechanisms are about equal; and above 45 km, the HO_x reactions control.

It appears that between 15 and 35 km, the oxides of nitrogen are by far the most important agent for maintaining the natural ozone balance. Calculations show that the natural NO_x should be about 4×10^9

molecules/cm^3. The extent to which this concentration would be modified by man-made sources such as supersonic aircraft determines the extent of the danger to the normal ozone balance. It must be stressed that this question is a complex one since both concentration and distribution are involved (see Johnston and Whitten, 1973).

3. The ClO_x Catalytic Cycle

Molina and Rowland (1974) pointed out that fluorocarbons could diffuse into the stratosphere and also act as a potential sink for ozone. Cicerone *et al.* (1974), show that the effect of these man-made chemicals could last for decades. This possible major source of atmospheric contamination arises because of the use of fluorocarbons as propellants and refrigerants. Approximately 80% of all fluorocarbons released to the atmosphere come from these sources. There is no natural source. 85% of all fluorocarbons are F11 (CCl_3F) or F12 (CCl_2F_2).

According to Molina and Rowland (1974), these fluorocarbons are removed from the stratosphere by photolysis above altitudes of 25 km. The primary reactions are

$$CCl_3F + h\nu \longrightarrow CCl_2F + Cl \qquad (159)$$

$$CCl_2F_2 + h\nu \longrightarrow CClF_2 + Cl \qquad (160)$$

Subsequent chemistry leads to release of additional chlorine, and for purposes here it is assumed that all of the available chlorine is eventually liberated to form compounds such as HCl, ClO, ClO_2, and Cl_2.

The catalytic chain for ozone which develops is

$$Cl + O_3 \longrightarrow ClO + O_2 \qquad (161)$$

$$ClO + O \longrightarrow Cl + O_2 \qquad (162)$$

$$\text{net} \quad O_3 + O \longrightarrow O_2 + O_2$$

Other reactions which are important in affecting the chain are

$$OH + HO_2 \longrightarrow H_2O + O_2 \qquad (163)$$

$$Cl + HO_2 \longrightarrow HCl + O_2 \qquad (164)$$

$$ClO + NO \longrightarrow Cl + NO_2 \qquad (165)$$

$$ClO + O_3 \longrightarrow ClO_2 + O_2 \qquad (166)$$

$$Cl + CH_4 \longrightarrow HCl + CH_3 \qquad (167)$$

$$Cl + NO_2 + M \longrightarrow ClNO_2 + M \qquad (168)$$

$$ClNO_2 + O \longrightarrow ClNO + O_2 \quad (169)$$

$$ClNO_2 + h\nu \longrightarrow ClNO + O \quad (170)$$

$$ClO + NO_2 + M \longrightarrow ClONO_2 + M \quad (171)$$

The unique problem that arises here is that F11 and F12 are relatively inert chemically and have no natural sources or sinks such as CCl_4 does. The lifetimes of these fluorocarbons are controlled by diffusion to the stratosphere where photodissociation takes place as designated by reactions (159) and (160). Lifetimes of halogen species in the atmosphere have been given in "Fluorocarbons and the Environment" (1975). These values are reproduced in Table 5. The incredibly long lifetimes of F11 and F12 and

TABLE 5

Residence time of halocarbons in the troposphere[a]

Halocarbon	Average residence time in years
Chloroform ($CHCl_3$)	0.19
Methylene chloride (CH_3Cl_2)	0.30
Methyl chloride (CH_3Cl)	0.37
1,1,1 Trichloroethane (CH_3CCl_3)	1.1
^{12}F	330 or more
Carbon tetrachloride (CCl_4)	330 or more
^{11}F	1000 or more

[a] Based on reaction with OH radicals.

their gradual diffusion into the stratosphere pose the problem. Even if use of these materials were stopped today, their effects are likely to be felt for decades.

Some recent results indicate, however, that the rate of reaction (171) may be much greater than initially thought. If so, the depletion of ClO by this route could reduce the effectiveness of this catalytic cycle in reducing the O_3 concentration in the stratosphere.

REFERENCES

Bachmeier, F., Eberius, K. H., and Just, Th. (1973). *Combust. Sci. Technol.* **7**, 77.
Barrett, R. E., Hummell, J. D., and Reid, W. T. (1966). *Trans. ASME J. Eng. Power* **88A**, 165.
Berry, A. G. U., and Edgeworth-Johnstone, R. (1944). *Ind. Eng. Chem.* **36**, 1140.
Blacet, F. R. (1952). *Ind. Eng. Chem.* **44**, 1339.

Bowman, C. T. (1954). *Int. Symp. Combust., 14th* p. 729. Combustion Inst., Pittsburgh, Pennsylvania.

Bowman, C. T. (1975). *Progr. Energy Combust. Sci.* **1**, 33.

Bowman, C. T., and Seery, D. V. (1972). "Emissions from Continuous Combustion Systems," p. 123. Plenum Press, New York.

Bracco, F. O. (1973). *Int. Symp. Combust., 14th* p. 831. Combustion Inst., Pittsburgh, Pennsylvania.

Bulfalini, M. (1971). *Environ. Sci. Technol.* **8**, 685.

Bradley, J. N. (1969). "Flame and Combustion Phenomena," Chap. 7. Methuen, London.

Chapman, S. (1930). *Phil. Mag.* **10**, 369.

Cicerone, R. J., Stolarski, R. S., and Walters, S. (1974). *Science* **185**, 1165.

Clarke, A. E., Hunter, T. G., and Garner, F. H. (1946). *J. Inst. Petrol.* **32**, 313.

Crutzen, P. J. (1970). *Roy. Meteorol. Soc. Q. J.* **96**, 320.

Cullis, C. F., and Mulcahy, M. F. R. (1972). *Combust. Flame* **18**, 222.

Dainton, F. S., and Irvin, K. J. (1950). *Trans. Faraday Soc.* **46**, 374, 382.

de Soete, C. C. (1975). *Int. Symp. Combustion, 15th*, p. 1093. Combustion Inst., Pittsburgh, Pennsylvania.

Dryer, F. L. (1975). Princeton Univ. Aerosp. and Mech. Sci. Rep. No. 1224.

Eberius, K. H., and Just, Th. (1973). Atmospheric Pollution by Jet Engines, AGARD Conf. Proc. No. 125, p. 16–1.

Fenimore, C. P. (1971). *Int. Symp. Combust., 13th* p. 373. Combustion Inst., Pittsburgh, Pennsylvania.

Fenimore, C. P. (1972). *Combust. Flame* **19**, 289.

Feugier, A. (1975). *Int. Symp. Chem. Reaction Dynam., 2nd* Univ. of Padua, Padua, Italy.

Flagan, R. C., Galant, S., and Appleton, J. P. (1974). *Combust. Flame* **22**, 299.

Fluorocarbons and the Environment (1975). Council on Environmental Quality, Washington, D.C.

Glassman, I., Dryer, F. L., and Cohen, R. (1975). *Int. Symp. Chem. Reaction Dynam., 2nd* Univ. of Padua, Padua, Italy.

Gordon, A. S., Smith, S. R., and McNesby, J. R. (1959). *Int. Symp. Combust., 7th* p. 317. Butterworths, London.

Haynes, B. S., Iverach, D., and Kirov, N. Y. (1975). *Int. Symp. Combustion, 15th*, p. 1103. Combustion Inst., Pittsburgh, Pennsylvania.

Homann, K. H., and Wagner, H. G. (1967). *Int. Symp. Combust., 11th* p. 371. Combustion Inst., Pittsburgh, Pennsylvania.

Hunt, B. G. (1965). *J. Atmos. Sci.* **23**, 88.

Hunt, B. G. (1966). *J. Geophys. Res.* **71**, 1385.

Johnston, G. M., Matthews, C. J., Smith, M. Y., and Williams, D. V. (1970). *Combust. Flame* **15**, 211.

Johnston, H. (1971). *Science* **173**, 517.

Johnstone, H., and Whitten, G. (1973). Atmospheric Pollution by Aircraft Engines, AGARD Conf. Proc. No. 125, p. 2–1.

Kesten, A. S. (1972). *Combust. Sci. Technol.* **6**, 115.

Leighton, P. A. (1961). "Photochemistry of Air Pollution," Chapter 10. Academic Press, New York.

Leonard, R. A., Plee, S. L., and Mellor, A. M. (1976). *Combust. Sci. Technol.* **14**, 183.

Levy, A., Merryman, E. L., and Reid, W. T. (1970. *Environ. Sci. Technol.* **4**, 653.

Martenay, P. J. (1970). *Combust. Sci. Technol.* **1**, 461.

Martin, G. B., and Berkau, E. E. (1971). AIChE Meeting, San Francisco, California.

Merryman, E. L., and Levy, A. (1971). *Int. Symp. Combust., 13th* p. 427. Combustion Inst., Pittsburgh, Pennsylvania.

Merryman, E. L., and Levy, A. (1974). *Int. Symp. Combust., 15th* p. 1073. Combustion Inst., Pittsburgh, Pennsylvania.

Molina, M. I., and Rowland, F. S. (1974). *Nature (London)* **249**, 810.

Mulvihill, J. N., and Phillips, L. F. (1975). *Int. Symp. Combustion, 15th,* p. 1113. Combustion Inst., Pittsburgh, Pennsylvania.

Myerson, A. L. (1974). *Int. Symp. Combust., 15th* p. 1085. Combustion Inst., Pittsburgh, Pennsylvania.

Myerson, A. L., Taylor, F. R., and Harst, P. L. (1957). *J. Chem. Phys.* **26**, 1309.

Palmer, H. B., and Cullis, C. F. (1965). "Chemistry and Physics of Carbon" (P. L. Walker, ed.), Vol. I, Chapter 5. Dekker, New York.

Sachyan, G. A., Gershenzen, Y. M., and Naltandyan, A. B. (1971). *Dokl. Akad. Nauk. SSR* **175**, 647.

Schalla, R. L., and Hibbard, R. R. (1959). NACA Rep. 1300, Chapter IX.

Weinberg, F. J. (1962). *Combust. Flame* **6**, 59.

Weinberg, F. J., and Lawton, J. (1969). "Electrical Aspects of Combustion," Chapter 7. Oxford Univ. Press, London and New York.

Wendt, J. O. L., and Ekmann, J. M. (1975). *Combust. Flame* **25**, 355.

Zeldovich, Ya. B. (1946). *Acta Physecochem. USSR* **21**, 577.

Chapter Nine

The Combustion of Coal

In order to achieve an understanding of the burning of coal, a natural solid fuel containing carbon, moisture, ash, and a large number of different hydrocarbons, it is best to consider first a combustion process which has not been discussed to this point. This process includes surface burning such as that which occurs when carbon graphite burns. The volatiles in coal contribute a substantial amount to the energy release process. They burn rapidly, however, compared to the solid fuel which remains. It is the burning rate of the remaining solids, substantially carbon that determines the efficiency of any coal process.

A. DIFFUSIONAL KINETICS

Consider, for example, the burning of carbon graphite. The volatility of carbon is so small that it is not likely that a vapor-phase diffusion flame could exist in the same manner in which droplet burning was considered previously. Further, there is the uncertainty as to the kinetic rates in this true surface burning phenomenon. The liquid hydrocarbon droplet flames produce gaseous fuels which react with gaseous oxidizers at high temperatures and thus at a rate appreciably faster than the gaseous diffusion rates existing in the problem.

In the first instance, surface oxidation kinetics such as those for carbon graphite cannot be considered to be fast compared to the rate of diffusion of oxygen to the surface. The temperature of the surface will determine these reaction rates, and this temperature is not always known a priori.

The problem of determining the burning rate of carbon graphite is more complex because the assumption that the chemical kinetic rates are very much faster than the diffusion rates cannot be made.

Consider, for example, a carbon surface burning in a concentration of oxygen in the free stream specified C_∞. The burning is at a steady mass rate. Then the concentration of oxygen at the surface is some value C_s. If the surface oxidation rate follows first-order kinetics as Frank-Kamenetskii (1955) assumed, then

$$G_{ox} = G_f/i = kC_s \tag{1}$$

where G is the flux in gm/sec cm^2, k is the heterogeneous specific reaction rate constant for surface oxidation in units reflecting a volume to surface area ratio as well, i.e., centimeters per second; and i is the mass stoichiometric index. However, the problem is that C_s is unknown. But one also knows that the consumption rate of oxygen must be equal to the rate of diffusion of oxygen to the surface. Thus if h_D is designated the overall convective mass transfer coefficient (conductance), then one can write

$$G_{ox} = kC_s = h_D(C_\infty - C_s) \tag{2}$$

What is sought is the mass burning rate in terms of C_∞. It follows then

$$kC_s = h_D C_\infty - h_D C_s \tag{3}$$

$$kC_s + h_D C_s = h_D C_\infty \tag{4}$$

$$C_s = \{h_D/(k + h_D)\}C_\infty \tag{5}$$

or

$$G_{ox} = [\{kh_D/(k + h_D)\}C_\infty] = KC_\infty \tag{6}$$

$$K = kh_D/(k + h_D) \tag{7}$$

$$1/K = (k + h_D)/kh_D = (1/h_D) + (1/k) \tag{8}$$

When the kinetic rates are large compared to the diffusion rates, $K = h_D$; when the diffusion rates are large compared to the kinetic rates, $K = k$. When $k \ll h_D$, $C_s \cong C_\infty$ from Eq. (5), thus

$$G_{ox} = kC_\infty \tag{9}$$

When $k \gg h_D$, Eq. (5) gives

$$C_s = (h_D/k)C_\infty \tag{10}$$

But since $k \gg h_D$, it follows from Eq. (10) that

$$C_s \ll C_\infty \tag{11}$$

This result permits one to write Eq. (2) as

$$G_{ox} = h_D(C_\infty - C_s) \cong h_D C_\infty \tag{12}$$

Consider the case of rapid kinetics, $k \gg h_D$, further. In terms of Eq. (6), or examining Eq. (12) in light of K,

$$K = h_D \tag{13}$$

Of course, Eq. (12) also gives us the mass burning rate of the fuel

$$G_f/i = G_{ox} = h_D C_\infty \tag{14}$$

h_D is the convective mass transfer coefficient for an unspecified geometry. For a given geometry, h_D would contain the appropriate boundary layer thickness or would have to be determined by some independent measurements which would give correlations from which h_D could be determined from other parameters of the system.

B. THE BURNING RATE OF CARBON

This situation, as discussed in the last section, is then very much like the droplet diffusion flame discussed previously. The oxygen concentration approaches zero at the flame front, except now the flame front is at the particle surface and there is no fuel volatility. Of course, the droplet flame discussed before had a specified spherical geometry and was in a quiescent atmosphere. Thus h_D must contain the transfer number term because the carbon oxide formed will diffuse away from the surface. However, for the diffusion-controlled case, there is no need to proceed through the conductance h_D, as the system developed earlier is superior.

Recall for the spherical symmetric case of particle burning in a quiescent atmosphere, one has

$$G_f = (D\rho/r_s)\ln(1 + B) \tag{15}$$

where D is the mass diffusivity, ρ the gaseous density, r_s the radius of the particle, and B the transfer number. The most convenient B in liquid droplet burning was

$$B_{oq} = (im_{0\infty} H + C_p(T_\infty - T_s))/L_v \tag{16}$$

since even though T_s was not known directly, $C_p(T_\infty - T_s)$ could always be considered much less than $im_{0\infty}H$ and ignored. Another form of B is

$$B_{f0} = (im_{0\infty} + m_{fs})/(1 - m_{fs}) \tag{17}$$

Indeed this form of B was required in order to determine T_s and m_{fs} with the use of the Clausius–Clapeyron equation. This latter form is not frequently used because m_{fs} is essentially an unknown in the problem and cannot be ignored as the T_s term was in Eq. (16). It is, of course, readily determined and necessary in determining G_f. However, look at the convenience in the current problem. Since there is no volatility of the fuel, $m_{fs} = 0$, and one has from Eq. (17)

$$B_{f0} = im_{0\infty} \tag{18}$$

Thus, for surface burning with fast kinetics, a very simple expression is obtained:

$$G_f = (D\rho/r_s)\ln(1 + im_{0\infty}) \tag{19}$$

Notice whereas in liquid droplet burning B was not explicitly known because T_s is an unknown, in the problem of heterogeneous burning with fast surface reaction kinetics, B takes the simple form of $im_{0\infty}$ and is known provided the mass stoichiometric coefficient i is known. For small values of $im_{0\infty}$, Eq. (19) becomes very similar in form to Eq. (14).

The value of i for carbon or for that matter any heterogeneous reaction is not readily evaluated.

If the final product of the carbon surface reaction is carbon dioxide, then

$$C + O_2 \longrightarrow CO_2$$

where i the stoichiometric coefficient has been defined as the number of grams of fuel that burns with one gram of oxidizer. Thus for this reaction, $i = 12/32$. From knowledge of the structure of CO_2,

$$O = C = O$$

it is not likely that CO_2 would form as a gaseous product on the surface. It is much more reasonable that CO would form and the correct value of i would be $i = 12/16$ as evaluated from

$$C + \tfrac{1}{2}O_2 \longrightarrow CO$$

Indeed, experimental evidence appears to confirm that near the surface the product is CO and that the conversion of CO to CO_2 takes place in the gas phase in a thin reaction (flame) zone surrounding the particle.

Figure 1 details some of the evidence as reported by Coffin and Brokaw (1957). These results confirm what was inferred in the previous paragraph that

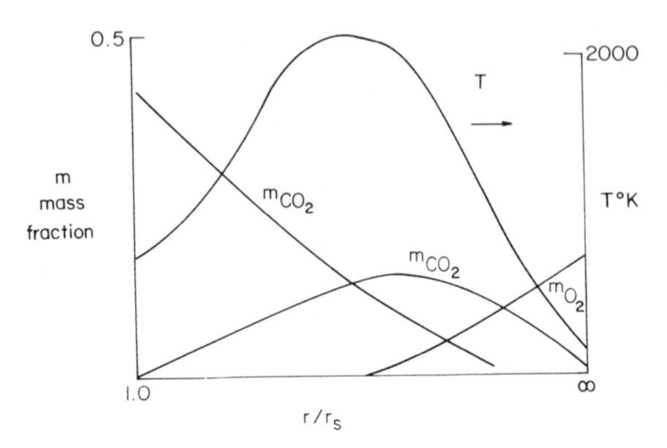

Fig. 1. Distribution of gases species and temperature above a carbon sphere burning on the surface (after Coffin and Brokaw, 1957).

carbon monoxide forms on the surface, diffuses away and reacts with the oxygen to form carbon dioxide. The carbon dioxide then diffuses in both directions—towards and away from the surface. When the CO_2 reaches the carbon surface, it is reduced to CO by the reaction

$$CO_2 + C \longrightarrow 2CO$$

Thus actually little, if any, oxygen reaches the carbon surface. What depletes the carbon is its reduction of CO_2 into CO. Yet the stoichiometric coefficient in B is related to the free stream mass fraction of oxygen. Since it is CO which forms at the surface, one would expect that the correct i is 12/16.

In order to verify that this value is the correct i for the sequence of reactions, one must proceed through the analytical development of the graphite particle burning. In this problem, because there is no combustion in the gas phase, one is required to deal only with the oxygen diffusion equation. Indeed this is the reason for the simple result that $B = im_{0\infty}$. The procedure of Blackshear (1960) is followed.

The mass diffusion equation developed earlier for droplet evaporation alone

$$4\pi r^2 \rho v \, dm_{ox}/dr = (d/dr)(4\pi r^2 \rho D \, dm_{ox}/dr) \tag{20}$$

now holds for the case where there is combustion on the surface, i.e., there is no reaction rate term in Eq. (20). In the gas phase i' grams of CO react with one gram of O_2 to give $(1 + i')$ grams of CO_2, therefore,

$$m_{O_2} = -[1/(1 + i')]m_{CO_2} \tag{21}$$

A new variable, physically symbolizing and allowing for the diffusing of both the oxygen and CO_2,

$$m_H = m_{O_2} + [1/(1+i')]m_{CO_2} \tag{22}$$

satisfies the fundamental differential equation for diffusion:

$$\rho v \, dm_H/dr = (d/dr)(\rho D \, dm_H/dr) \tag{23}$$

When this equation is integrated and evaluated at $r = r_s$, one obtains

$$\text{const} = r_s^2[\rho_s v_s (m_H)_s - D_s \rho_s (dm_H/dr)_s] \tag{24}$$

$4\pi r_s^2 \rho_s v_s$ is the mass consumption rate of the fuel and is a constant. At $r = r_s$, $m_H = [1/(1+i')]m_{CO_2}$ since $m_{O_2} = 0$. Thus the term in brackets is the flux of m_H into the surface, but it is, of course, also $(1/1+i')$ times the negative flux of CO_2 into the surface as well. But, by realizing that the statement that 1 gm of CO_2 can combine with i'' gm of C to form $(1+i'')$ gm of CO can be written, then the flux of CO_2 can be written in terms of the flux of carbon. Thus the flux of CO_2 must be $(1/i'')$ times the flux of C. The flux of carbon comes from the basic consumption rate of the carbon and the form is very much like that of liquid droplet consumption, i.e.,

$$\rho_s v_s = \rho_s v_s (m_H)_s \tag{25}$$

which is correct for solid or gas as long as the consistent value of density or velocity is chosen. Then the flux of CO_2 must be

$$\rho_s v_s (1/i'') \tag{26}$$

But the term in the brackets is $(1/1+i')$ times the flux of CO_2. Then the term in the brackets is

$$-\rho_s v_s (1/i'')[1/(1+i')] \tag{27}$$

and the equation is written

$$\text{const} = -r_s^2 \rho_s v_s (1/i'')[1/(1+i')] \tag{28}$$

Integrating the second time, one has

$$\frac{r_s^2 \rho_s v_s}{rD\rho} = \ln\left(\frac{m_{H\infty} + [1/i''(1+i')]}{m_H + [1/i''(1+i)]}\right) \tag{29}$$

However both CO_2 and O_2 must approach zero at $r = r_s$, therefore $m_H = 0$ at $r = r_s$ and, of course, $m_H = m_{O_2\infty}$ at $r = \infty$. Therefore

$$r_s \rho_s v_s/\rho D = \ln[m_{O_2\infty} i''(1+i') + 1] \tag{30}$$

Comparing this equation with the many forms that were obtained previously, one then has that

$$B = i''(1 + i')m_{O_2\infty} \tag{31}$$

The values of i' and i'' are $i'' = 12/44$ and $i' = 28/16$. Then

$$i''(1 + i') = (12/44)(1 + 28/16) = (12/44)(44/16) = 12/16 \qquad \text{or}$$

$$i''(1 + i') = 12/16 = i \tag{32}$$

Thus irrespective of the mechanism of removing carbon from the surface, only CO forms on the surface and the flux of oxygen from the quiescent atmosphere must be stoichiometric with respect to CO formation regardless of the intermediate reactions which take place. The same point can be seen by simple addition of the two primary reactions considered

$$\begin{array}{lcl} CO + \frac{1}{2}O_2 & \longrightarrow & CO_2 \\ CO_2 + C & \longrightarrow & 2CO \\ \hline C + \frac{1}{2}O_2 & \longrightarrow & CO \end{array}$$

Recall this discussion has been for $k \gg h_D$, which in the context of combustion reactions means high temperature particles. Of course, such high temperatures are created at the surface by accelerating the mass burning rate by increasing the convective rates. The convective expression for the burning of the carbon particle is of the same form as that of the burning liquid droplet, except that the expression for B is simpler in this case.

Various investigators have shown that the combustion of carbon above 1200°K exhibits rates which are strictly proportional to the diffusional characteristics.† The work of Dryer *et al.* (1971) shows that above 1100°K CO oxidation to CO_2 is very rapid. Thus it would appear that all the assumptions made are self consistent. If the CO oxidation rates were not rapid, one would have to be concerned with the oxygen penetrating to the carbon surface, and the overall analysis would become more complex algebraically, but the same result obtained nevertheless.

C. THE BURNING OF POROUS CHARS

Real coal particles have pores and are thus not like the ideal carbon particle discussed in the last section. Indeed one could simply look at the pore situation by saying that the pores give increased surface area to the

† For pulverized particles of the order of tens of microns or less, diffusional rates may be faster than chemical rates at temperatures in the 1000–2000°K range and reaction kinetics can be the controlling process (see Mulcahy and Smith, 1969).

carbon particle. Of course, if the diffusion rates are controlling, then the surface area of the particles does not play a significant role. What does play a role is the rate at which one can get the oxygen to the surface; i.e., the molecular or convective diffusion rate.

Physically it is better to consider a large surface area particle as one that has a great deal of pores. However, one can use physical arguments to distinguish between ranges of applicability when there are deep pores and a large, rough external surface area. Following the pore concept, it must be realized that there is penetration of oxygen into these pores, and the reaction or depletion of carbon takes place within these pores as well.

Consider now, as Knorre *et al.* (1968) have, the situation in which diffusion to the particle is sufficiently fast that it is not the controlling rate. For the porous medium, carbon is being consumed within the pores as well as on the surface. The surface consumption rates are therefore controlled by the kinetic rates; however, the consumption rates in the pores are controlled by the diffusion of oxygen into the pore. Thus the mass consumption rate of oxygen in terms of a flux of oxygen must be that which is consumed at the surface and that which diffused into the pores

$$G_{ox} = k(C_{O_2})_s + D_i(\partial C_{i,O_2}/\partial n)_s \tag{33}$$

where D_i is the internal diffusion coefficient and C_{i,O_2} the oxygen concentration within the particle.

Following a convention established earlier, this equation is written in the form

$$G_{ox} = k(C_{O_2})_s + D_i(\partial C_{i,O_2}/\partial n)_s = k'(C_{O_2})_s \tag{34}$$

The easiest manner to consider the problem is that the pore is one of spherical symmetry and thus the internal oxygen diffusion process is described by the equation

$$D_i\left(\frac{d^2C_i}{dr^2} + \frac{2}{r}\frac{dC_i}{dr}\right) - q_{O_2} = 0 \tag{35}$$

where q_{O_2} is the oxygen requirement (consumption) rate per unit particle volume.

It is possible to express the quantity q_{O_2} as

$$q_{O_2} = kS_i C_i \tag{36}$$

where S_i is the internal surface area in a unit particle volume m^2/m^3 which for a very porous particle would approximate the total surface area per unit volume.

The solution of Eq. (35) in terms of the expression as given in Eq. (34) results in the following expression for k':

$$k' = k + \lambda D_i[\coth(\lambda R) - 1/\lambda R] \tag{37}$$

where R is the particle radius and

$$\lambda = (S_i k)^{1/2}/D_i \tag{38}$$

For the case of small values of λR ($\lambda R < 0.55$) which physically is representative of burning at low temperatures or the burning of small particles, the $\coth(\lambda R)$ can be expanded into a series in which only the first two terms can be considered significant, i.e.,

$$\coth(\lambda R) \cong (1/\lambda R) + (\lambda R/3) \tag{39}$$

Substituting Eq. (39) in (37), one obtains

$$k' = k\{1 + (S_i R/3)\} \tag{40}$$

This expression then is the rate constant when the inner surface of the pores participate.

If the second term in the parenthesis of Eq. (40) is small compared to unity, i.e., if the internal surface area is small with respect to the external surface area of the particle, then

$$k' = k \tag{41}$$

Since S_i can be a very large number, tens of thousands of m^2/m^3, the condition $(S_i R/3) < 1$ may be satisfied only for very small particles which have radii of the order of tens of microns. Physically one would not expect that for very small particles there would be a large internal surface area compared to an external surface.

For large values of λR which correspond to high temperatures and large particles

$$\coth(\lambda R) \approx 1 \tag{42}$$

and Eq. (37) becomes

$$k' = k + (S_i Dk)^{1/2} \tag{43}$$

As the temperature increases, the first term in Eq. (43) increases more rapidly than the second because the temperature dependency is only in k. Therefore at high temperatures

$$k' = k \tag{44}$$

which simply means that the oxygen is completely consumed at the external surface. For moderate temperatures, it is found that

$$k' \approx (S_i D_i k)^{1/2} \tag{45}$$

or

$$k'/k \approx (S_i D_i/k)^{1/2} \tag{46}$$

$S_i D_i/k$ is a dimensionless number which arises in diffusional kinetics problems. In reality the best form for k' is Eq. (45) since Eq. (34) may now be written as

$$G_{ox} \approx (S_i D_i k)^{1/2}(C_{O_2})_s \approx (S_i D_i k)^{1/2} C_\infty \tag{47}$$

because external diffusion is fast and thus $(C_{O_2})_s \approx C_\infty$.

To this point the limit conditions have been handled but one can make some experimental headway by recalling Eq. (6),

$$G_{ox} = (kh_D/(k + h_D))C_\infty = KC_\infty \tag{48}$$

and Eq. (7)

$$K = kh_D/(k + h_D) \tag{49}$$

But, more generally for the possible porous problem, Eq. (49) can be written as

$$K = k'h_D/(k' + h_D)$$

where k' can take the values

$k' = k(1 + (S_i R/3))$	low temperature or small particles
$= k$(high temperature)	small particles
$= k + (S_i D_i k)^{1/2}$	high temperatures and large porous particles
$= (S_i D_i k)^{1/2}$	moderate temperatures and large porous particles

according to the various limit conditions defined.

D. THE BURNING RATE OF ASH-FORMING COAL

Some coals contain an ash in addition to carbon, moisture, and volatiles. To obtain a conservative estimate one should assume that a porous ash shell is retained during the burning of the combustible material. This ash may, of course, have a catalytic effect on the heterogeneous carbon combustion reactions but it is a cause, however, for additional diffusion resistance.

It is apparent that this shell offers great resistance to oxygen diffusion from the free stream to reach the combustible material. It does not matter whether the oxygen actually diffuses to the carboneous surface. The actual mechanism by which the carboneous material is consumed is probably very much like that for the pure carbon particle except that the CO to CO_2 conversion is largely heterogeneous. In this problem, diffusion of oxygen to the particle is very much faster than diffusion through the ash; hence, one can assume that the oxygen concentration at the edge of the shell is the same as the atmospheric concentration. The oxygen (or CO_2) concentration at the fuel surface approaches zero.

As a first approximation and realizing from the previous discussion that it does not matter whether O_2 or CO_2 gets to the surface, one can assume that a simple linear oxygen gradient determines the oxygen flux. Since it is not a convective problem, one can write a simplified expression as

$$G_{\mathrm{ox}} = D_{\mathrm{i}}(C_{\mathrm{O}_2,\infty}/x) \tag{50}$$

where x is the thickness of the ash as shown in Fig. 2, and D_{i} is a diffusion coefficient through the ash.

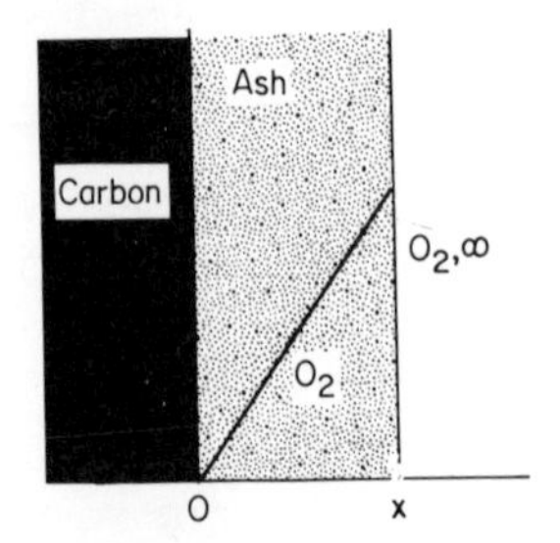

Fig. 2. Thickness of ash shell in diffusion combustion of high-ash fuel.

The fuel recedes at a rate dx/dt and is related to the oxygen flux by the expression

$$G_{\mathrm{O}_2} = \gamma\, dx/dt \tag{51}$$

where the density and stoichiometry are in γ. Thus at any given instant

$$\gamma\, dx/dt = D_{\mathrm{i}} C_{\mathrm{O}_2,\infty}/x \tag{52}$$

If one takes $x = 0$ at $t = 0$, the solution of Eq. (52) is

$$x = ((2D_{\mathrm{i}} C_{\mathrm{O}_2,\infty}/\gamma)t)^{1/2} \tag{53}$$

Substituting in Eq. (50)

$$G_{\mathrm{O}_2} = (D_{\mathrm{i}} C_{\mathrm{O}_2,\infty}\gamma/2)^{1/2}(1/\sqrt{t}) \tag{54}$$

or

$$G_{\mathrm{f}} \sim C_{\mathrm{O}_2,\infty}^{1/2}$$

It is not such a surprising solution that the oxygen flux decreases or that the consumption of fuel decreases with time as the ash thickness increases. What one obtains, however, is the inverse square root dependency with time and the square root dependency with concentration. Thus for an ash-forming fuel in which ash remains firm throughout the combustion process, the burning rate is proportional to the square root of the oxygen concentration and is independent of the convective nature of the oxidizer stream.

For nonash-forming coals, the burning rate according to Eq. (19) is

$$G_f \sim \ln(1 + im_{O_2}) \tag{55}$$

However, the mass fraction of oxygen in air is 0.23 and i is 0.75, and the product is 0.172. Thus im_{O_2} is small compared to 1 and

$$G_f \sim m_{O_2,\infty} \sim C_{O_2,\infty} \tag{56}$$

i.e., the burning rate is directly proportional to the oxygen concentration.

REFERENCES

Blackshear, P. L. Jr. (1960). "An Introduction to Combustion," Chapter V. Dept. of Mech. Eng., Univ. of Minnesota, Minneapolis, Minnesota.

Coffin, K. P., and Brokaw, R. S. (1957). NACA Tech. Note 3929.

Dryer, F. L., Naegeli, D. W., and Glassman, I. (1971). *Combust. Flame* **17**, 270.

Frank-Kamenetskii, D. A. (1955). "Diffusion and Heat Exchange in Chemical Kinetics," Chapter II. Princeton Univ. Press, Princeton, New Jersey.

Knorre, G. F., Aref'yev, K. M., and Blokh, A. G. (1968). Theory of Combustion Processes, Translated by Foreign Tech. Div., Wright-Patterson AFB, Ohio, FTD-HT-23-495-68 (2 vols), Chapter 24.

Mulcahy, M. F. R., and Smith, I. W. (1969). *Rev. Pure Appl. Chem.* **19**, 81.

Index

I

K

L

M

N

O

A
B 7
C 8
D 9
E 0
F 1
G 2
H 3
I 4
J 5